CONFRONTING CLIMATE CHANGE

The Policy and Practice of Governance

SERIES EDITOR
Maria Ivanova

Published in association with the School of Public Policy and Urban Affairs, Northeastern University

CONFRONTING CLIMATE CHANGE

From Mitigation to Adaptation

John Barkdull

LYNNE RIENNER PUBLISHERS
BOULDER
LONDON

Published in the United States of America in 2024 by
Lynne Rienner Publishers, Inc.
1800 30th Street, Suite 314, Boulder, Colorado 80301
www.rienner.com

and in the United Kingdom by
Lynne Rienner Publishers, Inc.
1 Bedford Row, London WC1R 4BU
www.eurospanbookstore.com/rienner

Library of Congress Cataloging-in-Publication Data
Names: Barkdull, John, 1954– author.
Title: Confronting climate change : from mitigation to adaptation / John Barkdull.
Description: Boulder, Colorado : Lynne Rienner Publishers, Inc., [2024] | Series: The policy and practice of governance | Includesibliographical references and index. | Summary: "Assessing more than three decades of climate policies and politics, Barkdull explores approaches to transformational adaptation that will allow us to survive in a much warmer world"— Provided by publisher.
Identifiers: LCCN 2024013817 (print) | LCCN 2024013818 (ebook) | ISBN 9781962551212 (hardcover ; acid-free paper) | ISBN 9781962551229 (ebook)
Subjects: LCSH: Global environmental change. | Climatic changes—Social aspects. | Climate change mitigation. | Ecosystem management. | United Nations Framework Convention on Climate Change (1992 May 9). Protocols, etc. (2015 December 12)
Classification: LCC GE149 .B365 2024 (print) | LCC GE149 (ebook) | DDC 363.7/07—dc23/eng/20240615
LC record available at https://lccn.loc.gov/2024013817
LC ebook record available at https://lccn.loc.gov/2024013818

British Cataloguing in Publication Data
A Cataloguing in Publication record for this book is available from the British Library.

Printed and bound in the United States of America

∞ The paper used in this publication meets the requirements of the American National Standard for Permanence of Paper for Printed Library Materials Z39.48-1992.

5 4 3 2 1

Contents

Acknowledgments

I WISH TO EXPRESS MY APPRECIATION TO THOSE WHO SHAPED my thinking and helped me to carry out this project. For many years I discussed the issues raised in this book with Paul G. Harris of the Education University of Hong Kong, and I owe him a great debt of gratitude for his insight and knowledge. I am grateful to the Texas Tech University Department of Political Science and former chair Timothy Nokken for funding, support, and encouragement as I conducted my research. I also appreciate the guidance on climate science provided by Katharine Hayhoe of the Texas Tech Climate Science Center and the Nature Conservancy, as well as Tom Rohrig's invaluable research assistance. My thanks also go out to the anonymous reviewers of the manuscript for this book and to the community of scholars researching and writing about one of the most significant topics in today's world. Importantly, I thank Marie-Claire Antoine of Lynne Rienner Publishers for her unfailing enthusiasm for the project, her patience, and her encouragement. Without her this book would not have been published.

1

The Challenge of Climate Change

CLIMATE SCIENTISTS TOLD US DECADES AGO THAT TEMPERatures were rising due to human activities, and that this would have significant ecological and social effects.[1] In its most recent Assessment Report (AR6), the Intergovernmental Panel on Climate Change (IPCC) asserted, "Human activities, principally through emissions of greenhouse gases, have unequivocally caused global warming, with global surface temperature reaching 1.1°C above 1850–1900 in 2011–2020."[2] Consequently, states the IPCC, odds of achieving the internationally agreed ambition to hold temperature rise below 1.5°C are low. Instead, current policy commitments would lead to global warming ranging from 2.1°C to 3.4°C by 2100, with 2.8°C being most likely. Moreover, AR6 asserts, there is an "implementation gap" between commitments and actual practices, meaning that global temperature will more likely reach 3.2°C above the baseline by the end of this century. The five-year Global Stocktake indicates that commitments countries have recently made could bring the expected temperature rise down to the 2.4–2.6°C range, but still sees significant shortcomings. Emissions remain too high to meet the Paris goals, and "adaptation efforts to date have focused on planning and have not yet driven the broad changes necessary to enhance adaptive capacity, strengthen resilience and reduce vulnerability."[3] Thus, despite some advances, global climate policy remains inadequate in its most important dimensions, and the systems transformations that various observers assert are required to redress this situation are remote.

International negotiations to address climate change began in the late 1980s, and with every round the sense of urgency has increased, but without sufficient policy responses. In 2007, a member of the IPCC declared that failure to act before 2012 would mean the world had waited too long, that the coming two or three years were a "defining moment" for climate policy.[4] Over fifteen years later, the defining moments continue to pass, and the adequacy of climate action remains in doubt; the Secretary-General of the United Nations asserts that we are in an age of global boiling, on a highway to climate hell with our collective foot on the accelerator.[5] Indeed, any attention to the daily headlines informs the reader that rising temperatures and changing world climate are here.[6] Sadly, for the most part, scientific predictions have fallen short of subsequent observed changes.[7] The climate is changing faster and more profoundly than expected, more than at any time in human history. Thus, policymakers at global and national levels have mostly agreed (in principle, if not in practice) that to avert severe, dangerous effects on human society requires urgent measures to reduce the amount of warming along with equally urgent measures to prepare society for the consequences of climate change.

Accordingly, global policy formulated in such venues as the annual Conference of the Parties to the UN Framework Convention on Climate Change (COP) attempts to combine *mitigation* (reducing temperatures) with *adaptation* (various responses to soften the effects of higher temperatures). According to this policy framework, too little mitigation will mean climate changes will be so severe that sufficient adaptation is not possible. At the same time, current and future effects of climate change mean that some adaptation is already happening and more will be required. A comprehensive report notes that insufficient efforts to cut back on emissions will mean that some adaptation goals will be out of reach as the impacts increase with every increment of temperature increase. Accordingly, the world could require sufficient mitigation to maintain conditions for successful adaptation.[8] Unfortunately, neither element of climate policy has shown sufficient progress.

In the years following the 1992 Rio Earth Summit, the focus had been almost entirely on mitigation.[9] The first major agreement on climate policy, the UN Framework Convention on Climate Change (UNFCCC), negotiated at the Earth Summit, included references to adaptation, but its policy focus was squarely on reducing the sources of higher temperatures. However, as observed and future impacts were registered, adaptation rose on the global agenda, with demands for more funding, research, and capacity building. Since about 2007, high-level negotiations have seen claims that mitigation and adaptation are equally important, with both required for an adequate response to a significantly altered global climate.

Unfortunately for the mitigation + adaptation formula, global and national policies to reduce global warming below dangerous levels have shown insufficient signs of success so far, nor is there much hope for adequate reductions in the future.[10] Projections of temperature rise show that the world is on track to surpass internationally agreed thresholds within a few decades, with temperatures continuing to rise well beyond the end of this century. For decisionmakers, the evident inadequacies of mitigation policy have shifted their attention to how to live on a hotter planet. For the world, the upshot is that adaptation will be increasingly necessary, and the main question will be: How much social change will suffice to meet the challenge? That is, adapting to a much warmer world is primarily a social and political question, not a technological one.

The failure of global policy to reduce projected temperature rise sufficiently puts the onus on making major changes in the modern way of life to ameliorate the harms society will experience due to climate change.[11] Adaptation policy aims to reduce the vulnerability of society to such climate impacts as heat waves, extreme weather events, prolonged droughts, floods, and rising sea levels. It also seeks to increase *social resilience,* which is the capacity of society to recover from damages arising from climate change. But adaptation can only go so far. If temperatures rise too high, then the limits of social orders to which humans have become accustomed could be breached.[12] Although human extinction due to climate change is unlikely,[13] civilization as we know it could prove unviable, and profound social transformation would occur, whether by conscious design or as a chaotic, unplanned reaction to unprecedented climatic conditions.[14]

Can We Do It?

If adaptation to climate change is a kind of last-ditch defense against a social crisis resulting from human interference in the planet's climate, the question arises: Can we do it? Can we adapt enough and in the right ways to live in a hotter world without experiencing social chaos and the possible collapse of civilization? This is both a technical question and a political question. That is, we need to know what can be done on the ground to soften the blows of climate change. To what extent can flood protection, water conservation, changes in agricultural practices, forestry management, and the like reduce climate impacts to tolerable levels? We also need to know whether the institutions and processes for making and implementing climate policy can do the job.

Do they embody values, principles, norms, and prior commitments that prevent them from making the needed changes and transformations to sustain civilization and to achieve ambitious objectives such as fulfilling the promises of the Sustainable Development Goals (SDGs)? Do current institutions rely on decisionmaking processes and procedures that inhibit or block appropriate policies? If so, then what alternative institutions and practices can do the job? One finds in reviewing scholarly and official discourse on climate change adaptation an unfortunate tendency to project a dire—possibly catastrophic—future, but to balk at outlining what that future implies for the capacity of current institutional arrangements to consider, adopt, and implement adaptation measures sufficient to the task.

The IPCC, the entity that provides the UN with scientific reports on the state of the climate and possible policy responses, has said that rising temperatures might require reconsideration of current development pathways, and it has gone further in asserting in no uncertain terms that sustainable development in a warming world will require more democracy, equality, and social justice. These recommendations apply to adaptation as much as mitigation; the IPCC claims that injustice will hobble adaptation efforts, leaving vast numbers of people vulnerable to the worst effects of climate change.

The IPCC specifies criteria for successful adaptation, noting that adaptation solutions should be effective, feasible, and in conformity with principles of justice. Climate justice, writes the IPCC, encompasses distributive, procedural, and recognition justice. The first is about the allocation of costs and benefits from climate policy, the second is about who participates and decides, and the third is in regard to respect for and engagement with the diversity of cultures and perspectives.[15] Thus, the IPCC recommends adaptation that is fair to all stakeholders, employs legitimate decision processes, and guarantees the rights and participation of all. Adaptation that does not improve the condition of disadvantaged people is unsuccessful adaptation, which would, by the same token, stand as an indictment of the current state of the world, which is far from fair, democratic, or egalitarian. This would seem to call for a focused critique of existing global and national institutions, including the economic relationships producing vast inequality in wealth and access to decisionmaking processes. But the IPCC avoids tackling head-on the undemocratic, inequitable, unjust aspects of current global social, economic, and political institutions. Neither does it describe in detail the institutional framework that would enable successful adaptation to a considerably warmer climate.

To fill that gap requires turning attention to bodies of literature that the IPCC and other influential observers neglect, and to dissident voices outside official circles. The message coming from such sources is that sweeping social transformation is necessary to address all aspects of the climate challenge, including adapting to higher global temperatures. Whether such transformation can be enacted in time to head off social collapse is open to question. Indeed, one option is to accept that we are headed toward collapse and to prepare accordingly, which leads to consideration of such postcollapse proposals as Deep Adaptation.[16] In any case, questions of social organization become increasingly pressing, engaging issues regarding justice, equality, democracy, inclusiveness, class relations, and more.

Purpose of the Book

This book seeks to draw attention to the social and political requisites for successful adaptation to climate change. The orientation is global, focusing on international negotiations, international organizations, and the global implications of climate change. No doubt, much more could and should be said about national and local adaptation politics and policies, but this work views the issue through a global lens. This is appropriate to the present study because its central question is pitched at the broadest level: the future of civilization. Defending civilization is necessarily a collective global endeavor; this requires investigating, assessing, and evaluating research, political discourse, policy, and politics in international contexts.

The premise for the book is that efforts to reduce the sources of global warming—mainly the burning of fossil fuels—have failed and show little signs of future success. Thus, the world will heat up well beyond the internationally agreed thresholds for avoiding dangerous climate change. Consequently, adaptation to the tough new planet[17] brought about by climate change will become increasingly necessary.[18] Further, the degree of adaptation will become greater, requiring more extensive changes to how we live. In current terminology, sufficient, just adaptation to climate change might require "transformational change" of global institutions and practices. However, the barriers to transformational change can be as daunting as those to mitigation and adaptation.

As the social effects of climate change become more severe, the question will become increasingly urgent. Can current institutions implement adequate adaptation to defend the lives, property, and interests of more than a select few of the world's inhabitants, or are today's institutions barriers to the kind of democratic, equitable, just adaptation that the IPCC and others call for?[19] This book intends to address the institutional question directly, first through a chronological analysis of the development of adaptation policy and politics, then through consideration of alternative views of institutional change to meet the climate crisis. The goal is to reveal the significant turns in adaptation negotiations that have, so far, blunted the transformational potential of adaptation to climate change. This discussion points toward recent developments in adaptation thinking in academia and among activists to recover that potential. Toward that end, various options for transformational change will be elaborated.

The Challenge of Climate Change

Two decades ago, despite scientific consensus to the contrary, controversies still raged over whether human-caused temperature rise was even happening or, instead, observed temperature changes represented a cyclical phenomenon due to various natural causes. Some skeptics claimed that left-wing activists and self-interested scientists seeking funding were exaggerating a temporary, natural tendency in the planet's climate to impose a radical agenda of social change, while opponents argued that the science was honest and clear in projecting dangerous global warming would occur. That debate is over.

Today, the scientific consensus enjoys nearly universal assent: global average temperature is rising; human activities have caused the observed temperature increase and will induce even more in the future; higher global average temperature is already having observable effects on the ecosphere and will have greater effects as the world warms; human society will be significantly affected, mostly in negative ways; and the time to act is now. In its most recent Assessment Report (AR6), the IPCC asserts, "It is unequivocal that human influence has warmed the atmosphere, ocean and land. Widespread and rapid changes in the atmosphere, ocean, cryosphere and biosphere have occurred."[20]

The evidence is compelling, and opponents of policies restricting the use of fossil fuels and other mitigation policies have turned to advocating such measures as geoengineering and carbon capture. Indeed, fifty-two Republicans in the US Congress have formed a Conservative Climate Caucus to propose policies "to reduce emissions while retaining fossil fuels."[21] Achieving successful geoengineering would mean using massive technological interventions to cool the planet or to reinject greenhouse gases into the earth, obviating the need to cut back on burning fossil fuels and other human activities that raise global temperatures. The viability of such measures is questionable,[22] but unless various technological solutions succeed, not even the pledged goals of the 2015 Paris agreement can be met, not to speak of more ambitious objectives.[23] Yet deployment of such technologies on the scale required in time to make a significant difference faces serious challenges, and the unwanted and unplanned consequences could exceed the benefits.

Rising Temperatures

The fundamentals of the science of climate change have been known for over a century. The Earth's atmosphere contains certain gases that trap heat near the surface rather than allowing it to dissipate into space. Consequently, the planet maintains a global average temperature well above what it would be without those gases, and that means in turn climatic conditions compatible with life. In that respect, then, greenhouse gases are a blessing, without which the planet would be cold, dry, and dead. The problem, however, is that adding to the concentrations of greenhouse gases (GHG) from human sources can change the climate in unwanted ways, heating the planet to temperatures higher than ever experienced in the history of civilization.[24]

The most important of the atmospheric gases for the Earth's temperature is carbon dioxide (CO_2). When humans began to burn fossil fuels, releasing CO_2 in excess of the planet's capacity to absorb it quickly, the atmosphere retained more heat, raising global average temperatures. At first, this effect was barely detectable, but in the past several decades, a marked increase has occurred, so that global average temperature is now about 1.2°C higher than it was in preindustrial times,[25] and significantly higher in 2023. This is due to human activities increasing the concentration of GHG, primarily $CO_{2,}$ from about 280 parts per million of CO_2 and equivalents to 500 parts per million today.[26]

The world will heat further, with some warming already locked in, and some depending on which emissions pathway we take. Currently, the path is toward dangerous climate change as defined in international negotiations, specifically global average temperature rise higher than 2°C. Despite decades of negotiations to prevent global warming, global average temperature continues to rise. As José A. Tapia Granados observes, "Indeed, the sad reality is that besides bombastic and grandiloquent words, nothing has been done to stop the worsening of climate change since the process was discovered several decades ago."[27] Since the 1980s, every decade has been warmer than the previous one.[28] In 2021, the American West suffered through an intense heat wave that began before summer had even arrived. The European Union's Copernicus Programme reported that 2020 tied 2016 as the hottest on record, despite the cooling effects of La Niña.[29] In July 2023 the world experienced the hottest month on record even before the full effects of El Niño had been felt, and for all of 2023 global average temperature was 1.45°C above the preindustrial level, the hottest year on record.[30]

In 2011, the International Energy Agency (IEA) stated that the world had five years to change course on energy production or it would lock in too much global greenhouse emissions to avoid dangerous climate change.[31] The critical five years have long since passed without sufficient change in the global energy infrastructure. Indeed, the IEA reports that oil demand reached record highs in 2023.[32] "Climate and earth system science," writes Henrik Enroth, "make it clear to those willing and able to pick up the message that the time has long passed when anthropogenic climate change might have been reversed, and it presently appears that the window is closing on an opportunity to avert or at least reduce consequences of global warming that scientists agree will be nothing short of catastrophic."[33]

To be sure, most newly installed power generation has been nonfossil fuel, but that in itself only changes the mix of growing energy use, without necessarily reducing the amount of fossil fuels and GHG emissions. Rather, CO_2 emissions from fossil fuels have increased steadily for decades. While some regions have shown declines in this category of GHG, the world total reached 36.7 gigatons in 2020, up from 35 gigatons a decade earlier. Accounting for carbon dioxide and equivalents brought the figure to 50 gigatons in 2020, up about 30 percent from 1990.[34] In 2022, the amount of CO_2 and equivalents reached 59 gigatons. Yet to meet internationally agreed goals for containing temperature rise would require halving GHG emissions by 2030 and achieving net zero emissions by 2050. Assessing across forty indicators of progress on climate policy, an important study concludes that "none are on track to reach 2030 targets."[35]

The official world of international organizations, nongovernmental organizations, and governments has pinned its hopes on the 2015 Paris agreement on climate change, but the accord is likely to fail in its main aims. Parties to the Paris agreement officially adopted the goal of holding global average temperature rise below 2°C and, further, that they should pursue efforts to hold the rise below 1.5°C. The Paris agreement called on countries to make nationally determined contributions (NDCs) to reduce greenhouse gas emissions. But the NDC pledges to reduce GHG still leave the world on track for as much as 3°C global average temperature rise, with even higher levels possible.[36] Morgan Bazilian and Dolf Gielen observe that NDCs "don't come close to adding up to the Paris Agreement's goals," and existing commitments are not supported in national policies.[37] Without considerably increased commitments to cuts in GHG emissions and deployment of unproven carbon capture technology, global average temperature will rise well past the dangerous level. Indeed, the 2°C marker was arrived at politically and may represent what negotiators thought possible rather than an actual safe boundary; thus, we may already be in dangerous territory.[38]

Scientists have employed a variety of methods to estimate temperatures, including for times when no instruments to measure temperature existed. The data from various authorities show that the temperature trend turned significantly upward in the late 1960s, rising steadily since; estimates from six reputable scientific sources using different methods all report the same trend and nearly the same amount of change.[39] Certainly, fluctuations occur, but the fact that global average temperature is on a long-term upward trend is beyond dispute. Various alternative explanations have been forwarded for this trend (sunspots, natural cycles, and others), but the only one that survives serious scrutiny is that human activities—primarily burning fossil fuels—are responsible for the observed temperature rise and will be for future increases.[40]

Climate change's effects on ecosystems and societies have already been observed with the 1.2°C temperature rise. Some societies are now experiencing harms, and the trend is toward surpassing the thresholds for dangerous climate change, whether looked at in terms of what governments have promised to do, or in terms of what is actually happening in energy use and other activities contributing to global warming. The IPCC assesses the projected temperature ranges under four scenarios, ranging from a rapid reduction of greenhouse gas emissions to unrestricted growth. With rapid, steep reductions in fossil fuel use and other sources of global warming, it is possible to hold temperature rise to about where it is today. But less optimistic assessments of future patterns suggest the global average temperature will rise as much as 5°C.[41]

The path charted by the Paris agreement is toward about 2.8–3.0°C rise by the end of the century. To achieve the 2°C goal would require that countries fulfill their pledges to reduce GHG emissions, that commitments are "ratcheted up" in future negotiations, and that geoengineering technologies are developed and deployed broadly.[42] But practical actions to achieve pledges are so far insufficient. "A five-fold acceleration in emissions reductions is needed to come close," note Bazilian and Gielen.[43] For the United States, President Joe Biden's decision to rejoin the Paris agreement helped, but a major part of his climate policy was in the failed Build Back Better legislation, killed by coal-state Senator Joe Manchin (D-WV) along with all Senate Republicans.

Decades from now, the problem could become far worse. Although most depictions of the climate future show temperature increases by the year 2100, the end of this century is not the final chapter. Especially on the high-emissions path, warming and other effects of higher GHG concentrations continue for up to a millennium. The IPCC's Fifth Assessment Report (AR5) released in 2013 employed a variety of models to project future conditions.[44] The high-emissions model (called RCP8.5) shows global average surface temperature rising to about 4°C by 2100, then continuing to rise to 8°C in 2300. Although low probability, the temperature rise could reach 8°C by 2150 and climb to well over 12°C after 2250.[45] Importantly, in this process, surface warming will continue even after the factors forcing climate change have been eliminated or held constant. The long-term equilibrium surface temperature, then, is higher than the "transient" temperature; that is, what is being experienced at a given time. The emissions and concentrations created today will result in higher temperatures in the future, even if we immediately stop all emissions and other "forcings" entirely. This is because the concentrations of GHG in the atmosphere do not drop quickly once emissions stop. CO_2 in particular remains for centuries, so whatever level is reached before the world achieves net zero emissions is what determines long-term temperatures. In addition, the oceans hold massive amounts of heat, which has reduced the surface temperature rise that would otherwise have occurred. But that heat remains in the global climate system and will contribute to warming the planet. The National Oceanic and Atmospheric Agency observes that the heat absorbed by the oceans remains in the Earth system, it will eventually warm land areas, and "heat energy in the ocean can warm the planet for decades after it was absorbed."[46]

Thus, the policy horizon of 2100 generally employed to portray the climate future can obscure the continued warming of the planet beyond that time frame. Consequently, simply reducing the rate of growth of GHG emissions or reducing emissions by less than net zero might only delay matters.

Needless to say, a 12°C warmer world would have long since surpassed the point at which dangerous climate change is expected, and the survival of civilization would be unlikely. Paul G. Harris summarizes the current situation, "Emissions certainly are not set to fall by half within eight years, which scientists tell us is essential to limit global warming to 1.5°C and thus avoid the worst impacts of climate change."[47] Even worse, the World Meteorological Organization projects a strong chance of surpassing 1.5°C within this decade, a level nearly reached in 2023.[48] The trend continued into 2024, as April marked the eleventh straight month of being the hottest month on record, and the global average temperature from April 2023 to April 2024 was 1.61°C above preindustrial temperatures.[49] While the temperature average might temporarily fall below the 1.5°C threshold, the trend is clearly toward exceeding internationally agreed limits.

Finally, the annual report of the UN Secretary-General notes recent setbacks in climate policy, along with other elements of the Sustainable Development Goals. The report observes that, after a temporary reduction due to the Covid-19 pandemic, "global energy-related CO_2 emissions rose by 6% in 2021," and, further, existing commitments imply that "global emissions will increase by almost 14% over the current decade, leading to a climate catastrophe, unless governments and stakeholders" act immediately and collectively to head off the crisis.[50] But to hold global average temperature rise to 1.5°C "will require revolutionary scale transformational change in global social relations affecting the human relation to the climate and the planetary environment as a whole."[51] Thus, the Secretary-General's call for immediate collective action to head off a catastrophe is likely to go unheeded.

Effects of Climate Change

The IPCC concisely explains the observed and projected effects of climate change in its most recent report.[52] Already, writes the IPCC, "Human-induced climate change, including more frequent and intense extreme events, has caused widespread adverse impacts and related losses and damages to nature and people, beyond natural climate variability," and "the most vulnerable people and systems are observed to be disproportionately affected," pushing some social and natural systems "beyond their ability to adapt."[53] Climate and weather extremes have become more frequent and intense, leading to higher heat-related human mortality, coral bleaching, damage to forests, and storm damage. In addition, ocean acidification, sea level rise, and regional reduced precipitation have been attributed to human-induced global warming and burning of fossil fuels.

The ecological and social effects are already observed: "Climate change has caused substantial damages, and increasingly irreversible losses, in terrestrial, freshwater and coastal and open ocean marine ecosystems." Further, climate change has reduced food and water security, slowing the improvement of agriculture productivity and increasing malnutrition. Globally, climate change has had negative effects on agriculture, fisheries, the spread of infectious diseases, health, mental well-being, displacement from home communities, flooding, storm damages, infrastructure, and key economic sectors.[54]

The IPCC observes that the projected effects on social and ecological systems vary considerably due to "unsustainable development patterns" that leave 3.3 to 3.6 billion people highly vulnerable to climate change.[55] In the near term (2021–2040), when global average temperature is expected to reach and exceed 1.5°C permanently, global warming "would cause unavoidable increases in multiple climate hazards and present multiple risks to ecosystems and humans."[56] At the lower internationally agreed threshold, then, there will be more frequent and stronger extreme weather events, biodiversity loss, sea level rise encroaching on and even submerging coastal communities, and great stress on urban infrastructure. Social consequences include more conflict and displacement as people leave areas suffering severe climate change effects. High risks begin to emerge at 1.2°C (about where we were in 2023) and worsen as global average temperature rises by as much as 5°C compared to preindustrial times.

In the longer term, beyond 2040, key risks are "up to multiple times higher than currently observed."[57] Biodiversity loss could threaten nearly half the world's species at higher temperatures, and at the expected nearly ~3°C rise under current NDCs, as many as 29 percent of species would be threatened with extinction. The availability of water in regions relying on snow melt could fall by up to 20 percent at 2°C, and 40 percent at 4°C. At 2°C, the long-standing internationally agreed threshold for dangerous climate change, food security risks "will be more severe, leading to malnutrition and micro-nutrient deficiencies, concentrated in Sub-Saharan Africa, South Asia, Central and South America and Small Islands."[58] Sea level rise will expose as many as a billion people to climate effects specific to coastal areas, with additional damages occurring to communities and infrastructure in permafrost areas. Economic damages will increase, although too many uncertainties in the data exist to make confident assessments of a dollar amount. Involuntary migration, affecting the level of conflict, will increase as intense weather events, sea level rise, and lower agricultural productivity push people to seek new places to live. Finally, as temperatures rise, "multiple climate hazards will occur simultaneously, and multiple climatic and non-climatic risks will interact, resulting in compounding overall risks and risks cascading across sectors and regions."[59]

Summing up, climate scientists are increasingly confident of the consensus view that human activities have led to rising global average temperatures, with more increases to come over the next decades and centuries. Measurements from different sources employing different methods converge on the same pattern of rapid temperature rise over the past few decades, with nearly identical estimates of the amount of temperature increase. International agreements to address this challenge fall well short of what is required to avert dangerous climate change defined as no more than 2°C increase compared to preindustrial times. The more ambitious 1.5°C objective is all but out of reach, requiring improbable immediate reduction of GHG emissions, in turn meaning complete and rapid transformation of the global energy system.[60] Consequently, severe effects on ecological and human systems are likely to occur. The negative impacts multiply as temperatures rise, afflicting different regions and communities differently due, in part, to poverty, poor governance, and global inequality.

At higher temperatures, the capacity of modern civilization to cope is in doubt. At 4°C, Mark Lynas claims, "Advanced civilisation, with its constantly increasing levels of material consumption, energy use and living standards—the system that we call modernity and that has brought us into this crisis by continuing to be utterly dependent on fossil fuels—this civilisation is tottering."[61] Moreover, at such high temperatures relative to the preindustrial era, natural processes could overwhelm any human efforts to moderate further increases.[62] Consequently, adaptation to a much warmer world will continue to rise on the global agenda, and the degree of adaptation necessary will rise as well.

Adaptation Concepts

This section outlines key terms for the analysis of adaptation to climate change. To this end, it analyzes and relates adaptation, vulnerability, resilience, adaptive capacity, and transformational change.[63] The adaptation framework outlined here points to the essential task of specifying the institutional barriers to successful adaptation and the possibilities for institutional transformation to remove those barriers. But before discussing the adaptation framework, it is useful to elaborate the meaning of mitigation.

Mitigation

As noted, *mitigation* refers to efforts to prevent climate change from occurring. This means primarily reducing GHG emissions, with CO_2 the main target for reductions. However, mitigation extends beyond closing coal-fired power plants and the like. Other kinds of GHG and other factors also determine how warm the planet will become.

The IPCC identifies six major categories of sources of GHG emissions: electricity and heat production; agriculture, forestry, and other land use (AFOLU); buildings; transport; industry; and other energy. Electricity and heat production account for 23 percent of all GHG emissions, which the IPCC then breaks down to show the share for each of the other sectors, labeled indirect emissions. Combining the two (eliminating electricity and heat production as a separate category), AFOLU accounts for about 22 percent of all direct and indirect emissions, buildings for 16 percent, transport for 15 percent, industry for 34 percent, and other energy for 12 percent (Figure 1.1).[64]

Industry, transport, and buildings all rely heavily on direct or indirect use of fossil fuels. They also contribute other GHG such as methane. The cement industry accounts for a significant share of industry emissions. Mitigation means to reduce the GHG emissions from all of these sources.

In addition to reducing emissions, mitigation can be achieved by enhancing carbon sinks such as forests. Unfortunately, projections indicate that the most important forest in the world, the Amazon rainforest, could be headed toward a semiarid condition, even if excessive logging and land clearance were to cease.[65] Globally, the world is steadily losing forest cover, averaging 4.7 million hectares lost per year in the 2010s.[66]

Another approach is geoengineering, or the effort to employ massive technological remedies to the warming of the planet, without necessarily giving up fossil fuels and concrete or stemming deforestation. Broadly, geoengineering encompasses solar and carbon aspects. Solar geoengineering refers to efforts to limit the amount of sunlight reaching the earth's surface, while carbon geoengineering proposes using technological means to remove CO_2 from the atmosphere and inject it into the

Figure 1.1 Emissions by Source, 2019 (59 $GtCO_2$-eq.)

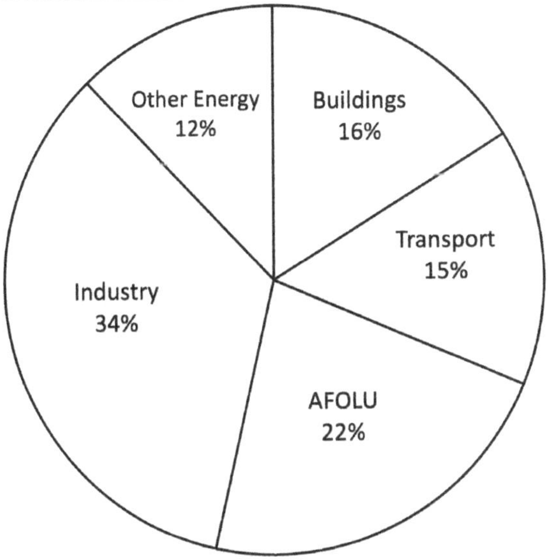

Source: IPCC. "Emissions Trends and Drivers," in *Climate Change 2022: Mitigation of Climate Change. Working Group III Contribution to the Sixth Assessment Report of the Intergovernmental Panel on Climate Change* (Geneva: IPCC, 2022), p. 237.

Note: AFOLU = agriculture, forestry, and other land use.

ground. Although the scale of a global project to control the planet's climate is immense, the Biden administration has sponsored carbon capture projects in Texas and Louisiana, among the latest of more than 1,200 projects worldwide, so the geoengineering ball is rolling.[67]

For the current commitments embodied in international agreements to be plausible, geoengineering must succeed. But considerable doubts about the efficacy of geoengineering exist, in part because the effects of climate control would fall differently on the various countries of the world. Those facing negative consequences or becoming relatively disadvantaged might not accept the outcome without resistance.[68] Moreover, numerous negative side effects are possible, serious enough that geoengineering to cool the planet could end up doing more harm than good, besides which the costs of building enough geoengineering infrastructure to make a difference could be prohibitive.[69] Indeed, the *Washington Post* reports that Occidental Petroleum has leased 106,000 acres of land near Corpus Christi just for the planned Texas carbon capture facility; installations around the world sufficient to matter would occupy large amounts of land.

In sum, mitigation encompasses emissions reductions in several important economic sectors, along with enhancement of carbon sinks and different forms of geoengineering. With various approaches to mitigation either falling well short or appearing infeasible, increasing attention turns to adaptation.

The Adaptation Framework

What is *adaptation* to climate change? Answering this question requires positing a definition and examining related concepts, thus enabling description and analysis of a framework for the study of adaptation.[70] This framework encompasses the sources of risk and vulnerability that necessitate adaptation; the capacities that households, communities, countries, and the world will need to develop to cope with climate threats; and the social transformations implied by adaptation to climate change. The climate challenge, from sources to effects to responses, is profoundly social. Adaptation is not about external natural causes imposing risks and costs on a distinct, separate social domain, mainly mandating technocratic solutions; it is about the social definition of what is at risk, what and who require protection, and what values are at stake in proposed responses to climate change. Certainly, technological and technical aspects exist, but vulnerability, developing adaptive capacity, achieving climate resilient development, and transformational change are inherently political, economic, social, and cultural phenomena.[71] Therefore, meanings and valuations vary by class, race, gender, and other identities.[72] Technocratic approaches, while possibly helpful in certain cases, will occur within a wider social and political context in which the meaning and significance of adaptation in all its dimensions is contested.

Table 1.1 displays definitions of adaptation from major policymaking bodies.[73] The IPCC, UN Environment Programme (UNEP), and UNFCCC see adaptation as "adjustments," while the UN Development Programme (UNDP) and UK Climate Impacts Programme (UKCIP) focus on "process," and the US government emphasizes actions to reduce risk. The common element is reducing the amount of harm from a warmer climate, while taking advantage of opportunities to benefit. Some adjustments will occur in response to experienced climatic changes such as shifting planting times as growing seasons reliably begin earlier in the year. Other adjustments can also take place in anticipation of future

Table 1.1 Definitions of Adaptation

Entity	Definition of Adaptation
Intergovernmental Panel on Climate Change (IPCC)	Adaptation, in response to current climate change, is reducing climate risks and vulnerability mostly via adjustment of existing systems. Climate-resilient development integrates adaptation measures and their enabling conditions with mitigation to advance sustainable development for all. Sixth Assessment Report (AR6)
UN Environment Programme (UNEP)	The process of adjustment to actual or expected climate and its effects. In human systems, adaptation seeks to moderate or avoid harm or exploit beneficial opportunities. In some natural systems, human intervention may facilitate adjustment to expected climate and its effects.
UN Framework Convention on Climate Change (UNFCCC)	Adaptation refers to adjustments in ecological, social, or economic systems in response to actual or expected climatic stimuli and their effects. It refers to changes in processes, practices, and structures to moderate potential damages or to benefit from opportunities associated with climate change.
UN Development Programme (UNDP)	Adaptation is a process by which strategies to moderate, cope with, and take advantage of the consequences of climatic events are enhanced, developed, or implemented.
UK Climate Impacts Programme (UKCIP)	Adaptation is the process or outcome of a process that leads to a reduction in harm or risk of harm, or realization of benefits, associated with climate variability and climate change.
US Fourth National Climate Assessment	Adaptation refers to actions taken at the individual, local, regional, and national levels to reduce risks from even today's changed climate conditions and to prepare for impacts from additional changes projected for the future. Adaptation is a form of risk management.

impacts such as strengthening building codes in expectation that extreme storms will occur more frequently. Regarding the process, the IPCC and various scholars have suggested establishing new policymaking bodies or inclusive decisionmaking procedures that pay greater attention to the experienced and anticipated impacts of climate change.

In the most recent IPCC report, adaptation is explained as follows:

> Adaptation plays a key role in reducing exposure and vulnerability to climate change. Adaptation in ecological systems includes autonomous adjustments through ecological and evolutionary processes. In human systems, adaptation can be anticipatory or reactive, as well as incremental and/or transformational. The latter changes the fundamental attributes of a social-ecological system in anticipation of climate change and its impacts. Adaptation is subject to hard and soft limits.[74]

Benjamin K. Sovacool and Björn-Ola Linnér identify four major areas for adaptation: infrastructural, institutional, community, and ecosystem. Respectively, they encompass such measures as building stronger water barriers, improving early warning systems, land use planning, and conservation.[75]

In addition to incremental and transformational adaptation, the IPCC has also suggested the possible need for transformational change. Transformational change addresses the institutional, ideological, and ethical contexts for adaptation decisionmaking, raising questions about economic and social institutions, as well as the beliefs and behaviors shaping a society's vulnerability and responsiveness to climate challenges. The IPCC asserts, "In cases where current development pathways are considered as the root causes of climate risk and vulnerability, transformation of wider political, economic, and social systems may be necessary."[76] It goes without saying that such profound change will disturb the status quo, including interests that benefit from existing arrangements.

Vulnerability

The degree to which adaptation is necessary depends on several factors, beginning with vulnerability.[77] An entity (individual, group, community, region, nation, the planet) faces climate risks to the extent it is vulnerable to adverse climatic changes. The IPCC defines *vulnerability* "as the propensity or predisposition to be adversely affected," encompassing "a variety of concepts and elements, including sensitivity or susceptibility to harm and lack of capacity to cope and adapt."[78] Vulnerability, then, is an attribute of the entity being affected. Something about its makeup renders it more or less likely to be adversely affected when some external factor changes. In the context of adaptation to climate change, as global temperatures rise, the more vulnerable communities will experience greater harms than those less vulnerable due to some set of attributes pertinent to those communities. Perhaps the most often cited factor to explain differences of vulnerability is poverty. Poor communities presumably have more weaknesses that will lead to greater harms from a given rise in global average temperature than a wealthier community would experience.

Adaptation policy tends to assume that the poor developing countries, many in Africa, are most at risk, requiring help from developed countries. This orientation can have the effect of rendering adaptation a technical matter in which experts guide investment in projects benevolently funded by wealthier countries to help the poor while avoiding questions of the systemic sources of the problem. Consequently, adaptation is understood mainly as "formal public sector interventions or programs," strengthening the role of policymakers and experts, shifting responsibility to local technical solutions "rather than in social struggles," marginalizing voices of the less privileged in the climate discourse.[79] Thus, adaptation policy in international negotiations tends to reproduce the parameters of the broader development agenda, in which the policymakers and experts are mainly from the Global North, providing guidance and technical advice to less capable leaders and communities in the Global South.

Vulnerability depends, in turn, on exposure. A society or community could harbor serious weaknesses in regard to climate change, but they would never manifest if there is no experience with the hazard or threat. To illustrate, a community might be ill-prepared to cope with high winds from extreme weather events, but if it never experiences them, its vulnerability will not become apparent.[80] The climatic effects of global warming vary considerably across locales, so the form that vulnerability takes varies as well. The effects of global warming will increase as temperatures rise. A community that enjoys low vulnerability today because the risk of exposure is low might not be so well situated in the future. Thus, vulnerability is a dynamic phenomenon requiring forward-thinking policy responses.

Further, vulnerability varies according to an affected entity's capacity to respond to a negative impact, which is termed resilience.[81] The IPCC defines *resilience* "as the capacity of social, economic, and ecosystems to cope with a hazardous event or trend or disturbance, responding or reorganising in ways that maintain their essential function, identity and structure as well as biodiversity in the case of ecosystems while also maintaining the capacity for adaptation, learning, and/or transformation."[82] In the natural world, this relates to an ecosystem's capacity to restore itself to its former state after some major disturbance. For example, a forest fire dramatically changes a region's ecology. In some cases, a predictable sequence of changes follows, restoring the forest to something like its previous state. But it could be that such a sequence never occurs and the ecosystem is permanently changed. The latter would be the less resilient, thus more vulnerable, forest. The same might apply to human societies, in which some can absorb a blow and return to some kind of preshock state, while others might be destroyed altogether. Enhancing resilience is an element of adaptation policy. Yet it is not certain that restoring a social system to its preshock condition is a desirable goal, which introduces debates over visions and values to the meaning of resilience.[83]

In sum, "In all formulations, the key parameters of vulnerability are the stress to which a system is exposed, its sensitivity, and its adaptive capacity."[84] Vulnerability, the IPCC notes, varies within and across regions due to "patterns of intersecting socio-economic development, unsustainable ocean and land use, inequity, marginalization, historical and ongoing patterns of inequity such as colonialism and governance."[85] Thus, vulnerability results from historical, social, political, economic, and cultural factors. As such, reducing vulnerability will require institutional remedies and is not merely a technical problem.

Adaptive Capacity

One element of reducing vulnerability and increasing resilience is to enhance adaptive capacity; this provides the link between vulnerability and adaptation.[86] *Adaptive capacity* is defined as "the property of a system to adjust its characteristics or behaviour, in order to expand its coping range under existing climate variability, or future climate conditions."[87] Adaptive capacity encompasses the resources needed for effective adaptation and the ability to mobilize those resources effectively. The latter includes both properly functioning policy formulation and implementation processes, and ability to overcome barriers to adaptation arising from external sources. In short, write Stephen R. Carpenter and William R. Brock, adaptive capacity is "the ability of a system to adjust to changing internal demands and external circumstances."[88]

Adaptive capacity is, in part, about gathering relevant information. This includes specification of the various social and ecological entities facing climate risks and hazards, and the options available for coping with these threats. Climate scientists, doctors, and engineers can use pertinent, quality information to recommend appropriate measures to reduce vulnerability and enhance resilience. The scholarly and policy literature has put considerable weight on the need for more and better research, and dissemination of the findings to the public. But adopting and implementing responses is a matter for policymakers in a political context. Accordingly, adaptive capacity means turning useful information into action, an uncertain process to be sure. If, for example, the IPCC's frequently stated view that inequality is a significant barrier to successful adaptation were to be taken seriously, then major changes would follow in social policy, taxation, income redistribution, wages, profits, and all the other elements of today's global civilization that have led to unprecedented wealth alongside deep poverty. But against entrenched interests, such changes are unlikely to occur. This would suggest that the adaptive capacity of current global institutions is low, no matter the quantity and quality of information about the problem. Needless to say, knowing that inequality undermines adaptive capacity has not led to extensive changes in institutions and practices.

Assessing adaptive capacity (and related dimensions of adaptation) requires first identifying the system of interest, whether at the local, regional, or global level.[89] This book is pitched primarily at the broadest level of concern; namely, civilization as we know it. Thus, the system of interest is of a global scale, and it has certain valued attributes that successful adaptation would preserve and enhance. Consequently, the question of whether today's prevailing institutions and practices possess sufficient adaptive capacity is, in some respects, about whether the global order is at risk and, if so, whether it can save itself.

Adaptive capacity requires key actors to accept that climate risks exist and that they are responsible for taking steps to cope with climate change effects. By the same token, contrary ideologies, vested interests, political ambitions, and other priorities can reduce adaptive capacity, leading to suboptimal adaptation policies. As Nick Brooks and W. Neil Adger observe, "Large scale structural economic factors and prevailing ideologies, therefore, play a vital role in determining which adaptations are feasible."[90] Thus, the major barriers to sufficient and effective adaptation lie in social and economic relationships and the ideas that support them.

What is the large-scale economic structure today, and what is the prevailing associated ideology? For now, suffice it to say that both structure and ideology can be captured in the word *neoliberalism*,[91] the predominant social and economic orientation of global and national affairs for the past four decades. Of course, the term *neoliberalism* is not synonymous with civilization, which existed before the current iteration of economic and social policy. Whether civilization as we know it can survive this neoliberal phase is at issue.

Do today's neoliberal structure and ideology have sufficient adaptive capacity and, thus, are they suited to long-term successful adaptation to climate change? If not, then transformational change of the global order would be required to meet the challenge. Significantly, adaptive capacity is about a system's ability to cope with external changes (a changing global climate, for instance) and to manage internal changes as well. Thus, questioning the adaptive capacity of today's prevailing institutions and ideologies is about more than their ability to cope with a changing climate. It is also about whether a system has the resources and capacities to change itself; that is, whether a system has the attributes and resources necessary to change internally so as to cope with the external challenge. On one hand, the power of entrenched interests, the weight of global institutions, and the pervasiveness of neoliberal ideology would suggest that today's social and economic order faces barriers too great to achieve internal transformation. Thus, the current ecological crisis might be beyond the adaptive capacity of any form of capitalist system. On the other hand, a broad historical perspective suggests that global capitalism has overcome limits on its development, including ecological limits, and might do so again.[92] Moreover, the fact that capitalism exists in a longer historical record indicates that broad social transformations can and do occur, and there is no reason to believe that capitalism is the last chapter in the story of humanity.

Transformation

Reducing vulnerability, enhancing resilience, and increasing adaptive capacity require political, social, economic, cultural, and institutional responses.[93] If today's global order lacks the adaptive capacity to prevent the collapse of civilization, then political, social, economic, and cultural institutions must look very different from current configurations. Consequently, calls for transformation are ubiquitous in the scholarly and policy literature on climate change. The IPCC's most recent report, for example, recommends transformational adaptation that will enable fulfillment of the Sustainable Development Goals.[94]

What constitutes the term *transformation?* Definitions vary widely, ranging from relatively small changes in a local community's decision-making processes to sweeping global changes in the institutions and values of the modern world. The meaning of transformation will be explored in depth in Chapter 4. Here, it is sufficient to note two contexts in which the IPCC has employed the term: transformational adaptation and transformational change.

Both incremental and transformational adaptation are about the measures taken to enhance resilience, reduce vulnerability, and increase adaptive capacity. Incremental and transformational adaptation lie on a continuum, "with no sharp division between them."[95] Nonetheless, useful distinctions exist. Incremental adaptation, which modifies existing systems without changing their basic features, refers to such actions as building a seawall to protect coastal installations as sea level rises. Transformational adaptation would encompass such measures as modifying land use regulations and instituting a managed retreat from the coast as sea level rises. Incremental adaptation maintains existing institutions and ways of life with marginal changes to limit the harms and take advantage of whatever opportunities climate change might create. It would not include significant redistribution of wealth and income, new forms of decisionmaking, or fundamentally different values. Transformational adaptation does call on societies to make significant changes such as relocating the community and changing economic pursuits. Together, incremental and transformational adaptation constitute the menu of options available to reduce social vulnerability to climate change. Selecting and implementing these forms of adaptation is contested: "A critical question then is how a better understanding of the political processes inherent in adaptation can help facilitate transformational adaptation, or alter the fundamental attributes and processes in society driving vulnerability (as well as rising emissions)."[96]

However, the IPCC recognizes that institutional factors might thwart both incremental and transformational adaptation measures.[97] Current development pathways, the IPCC says, could foreclose adaptation options that would limit adverse effects of climate change. Usually unstated in IPCC reports is that current development pathways are based on endless economic growth, markets, capital investment for profit, and the rest of the neoliberal prescription. Consequently, enabling appropriate, effective incremental and transformational adaptation could require fundamental reconfiguration of those institutions to the extent that they reduce society's capacity to respond. Institutional reforms that would enable making decisions and taking effective actions to put communities, countries, and the world on new development pathways is termed *transformational change*.

While avoiding a detailed description of the institutions, ideological paradigms, values, and development pathways standing in the way of both incremental and transformational adaptation, the IPCC does recognize that transforming them will be disruptive and divisive. In AR5, the IPCC wrote, "Transformational change can threaten vested interests, or prioritize the interests of some over the well-being of others, and it is never a neutral process, and it follows that transformation must be negotiated in the 'political sphere.'"[98] The concept of transformational change, while less prominent in AR6 than in AR5, provides an avenue to thinking about the aspects of prevalent global institutions that now and in the future will hinder adaptation to a much-changed climate. In other words, current institutions might face hard limits on their capacity for adaptation. To identify, choose, and implement necessary adaptation measures beyond the hard limits of existing institutions, then, could entail different institutions that do not face those hard limits. For example, if the commitment to endless economic growth driven by the profit motive is the institutional feature disabling appropriate incremental and transformational adaptation, then that commitment must give way to economic imperatives in line with climate adaptation needs; degrowth, for example.

In sum, the IPCC has employed the term *transformation* in two ways, one to refer to the measures taken to adapt society to a changed climate, the other to refer to the broader institutional framework within which decisions about adaptation are made. In most instances, incremental adaptations can be implemented without transforming social institutions, but that might not be true of transformational adaptations. As the climate crisis becomes more acute, the need for transformational adaptation will rise, and thus the adaptive capacity of current development pathways and institutions will come into question. Responding might require transformational change. Thus, the climate crisis poses the deepest questions regarding what kind of society we want.

Climate-Resilient Development

Building on a concept introduced in AR5, the IPCC lays greater stress in AR6 on climate-resilient development (CRD). Broadly, climate-resilient development seeks to integrate mitigation, adaptation, and sustainable development to enhance human and ecological well-being. The IPCC notes, "Current development pathways combined with the observed impacts of climate change, are leading away from, rather than towards, sustainable development."[99] Further, inequalities arising from gender, poverty, race, ethnicity, location, religion, and age increase vulnerability and reduce the adaptive capacity of societies and communities. Integrating various aspects of adaptation, the IPCC suggests that combining adaptation, mitigation, and sustainable development can make "all three more effective," as mitigation reduces the hazard while the other two reduce exposure and vulnerability. Adaptation and development work together to protect societies from climate change, as adaptation overcomes barriers to development while development expands the resources available to respond.[100] Choices made in the immediate future will determine future pathways, so time is growing short to transition to CRD.

Achieving CRD mandates rapid systems transitions in energy, industry, urban areas and infrastructure, land and ecosystems, and society. Such transitions, the IPCC asserts, "can widen the solution space and accelerate and deepen the implementation of sustainable development, adaptation, and mitigation actions by equipping actors and decision-makers with more effective options."[101] However, this will require considering a broad range of priorities, policies, and practices, moving beyond technical solutions and specific projects to giving attention to "the role of non-climate policies, social norms, lifestyles, power relationships and worldviews in enabling climate action and sustainable development."[102] Thus, AR6 is considerably more concerned than previous reports with questions of justice and equity. Yet what one does not find in the executive summary to the chapter on CRD, nor much anywhere else in the 3,675 pages of the full report, is elaboration of the characteristics of current development pathways blocking CRD, nor what would constitute a future post-transition society enabling CRD for all.[103]

The Stakes

What is at stake in the climate policy debate? Leaving aside those who deny that a problem even exists, the least alarmist view is that societies will have to adjust to serious, but manageable, effects of changing climate patterns, with those changes advancing gradually over the coming decades.[104] Some estimates of the economic impact of climate change, for instance, suggest that it could result in a reduction of potential world output by less than a few percentage points.[105] A small reduction in the growth of world output would not indicate the need for sweeping institutional reform. In the absence of significant economic effects, changes in other institutions and practices are likely to be incremental as well. Certainly, some regions and communities will experience greater impacts than others, but assistance can be provided to them while the broader social and economic framework continues as before.

However, many analysts project a much more dangerous future. For them, the observed effects of rising global temperature and expectations of higher temperatures in coming years indicate that the impacts will be disruptive, requiring major adaptation efforts in every region of the world and in every sector. Recent statements by climate scientists and other experts assert that a global crisis is coming unless rapid, deep reductions in GHG emissions are implemented along with many other measures to protect the ecosphere. An article signed by more than 11,000 scientists stated that the world needed to change its priorities from increasing economic output to sustainability, ensuring provision of basic needs to all people, and reducing inequality.[106]

Failing to do so, say some, could result in the collapse of civilization. Johan Rockström, a leading figure in research on sustainability, observes that we live on a small planet on which growth has become unsustainable and we are crashing through planetary boundaries, requiring a grand transformation if we are to remain within the planet's necessary conditions for civilization.[107] Transgressing planetary boundaries means that the Holocene conditions favorable to human civilization will no longer exist. The internationally agreed 2°C threshold presumably would provide tolerable conditions for the continuation of civilization more or less as we know it. Failing to stay somewhere near that limit as a maximum would produce a climate incompatible with modern society. This is not a prediction of human extinction, but it could suggest a return to the Stone Age. Some observers do consider extinction of the human species, along with many other species, to be a real possibility.

If the problem is grave, the response must match the challenge, meaning transformation of global institutions, which has been well and frequently expressed regarding mitigation. Targeting one aspect of the problem, Peter Dauvergne writes, "The unsustainability of global wealth creation is the underlying reason for the coming crisis of planetary instability."[108] Confronting the accumulation of wealth in a few hands would represent a revolutionary moment in world history, and that is only one aspect of the many social, economic, and cultural changes that would be required to achieve long-term sustainability. For another, the global ideology of consumerism would come into question. Leslie Sklair observes, that shopping, advertising and consumerism have been among the most successful and influential social forces in history."[109] Sklair further asserts that "the transnational capitalist class can be held responsible for ecological unsustainability because of its insatiable appetite for economic growth, and that, through the culture-ideology of consumerism, it exerts tremendous pressure on everyone on the planet towards consuming finite resources."[110] Consequently, if the crisis is to be severe enough to threaten civilization, then the response would demand challenging the growth imperative and its essential counterpart in the global ideology of consumerism. This is straightforward regarding mitigation, but adaptation too could call for significant institutional transformation.[111]

Theory and Climate Politics and Policy

How are we to understand adaptation to climate change theoretically? Approaches to the global politics of climate change adaptation encompass general theories of international relations (IR) and theories focusing on global environmental politics. The general theories—realism, liberal institutionalism, constructivism—tend to see the environment as an issue area like security and international political economy. Thus, the success or failure of adaptation to climate change would be understood as the outcome of the same forces, such as the balance of power, that explain wars, foreign trade, monetary relations, and the entire agenda of international relations. Other theories see unique attributes of environmental policy and politics, calling for theories that are explicitly ecological in orientation. For some, the environment is entwined with human society; theory should recognize the ecological basis for human society and that the unit of analysis is a socioecological system. The environment is not an issue area for political leaders to manage. Rather, it constitutes the social world.[112]

IR Theories

Influential general theories of international relations emphasize different aspects of global affairs: power, interests, institutions, cognitive factors, and class relations. Contending IR theories imply significantly different expectations regarding the future of climate policy and adaptation to climate change.[113]

Realists focus on power, arguing that the anarchy of the international political system requires states (the most important actors) to defend their security and survival against the military threats of other states. The upshot is that the balance of power explains behaviors and outcomes in world affairs. The rise of global environmental challenges changes little because security-seeking states must prioritize military preparedness and other enhancements of their power. When international cooperation does occur on nonsecurity issues, it is because powerful states cajole and pressure other states to abide by the norms and principles they favor in a given issue area. However, such collaboration is likely to be fleeting, swept away when the balance of power shifts to a less favorable configuration, or an international security crisis demands states' attention and efforts.[114] A realist point of view would thus suggest considerable skepticism that a decades-long global program to mitigate and adapt to climate change is feasible. Indeed, by 2100, we can expect that a major international security crisis, perhaps a world war, will have destroyed whatever climate policy framework states can manage to implement. If the war itself does not ruin civilization as we know it, the lack of measures to address climate change and the probability that the postwar climate situation would be even worse would finish the job. In any case, for realism, nothing changes in the essential character of world politics: the struggle for survival and power abides, and it thwarts effective responses to a plethora of social, economic, environmental, and political problems. Thus, realist theory would predict that global policy to mitigate and adapt to climate change will be, to the extent it requires international cooperation, weak, short-lived, and low priority compared to security and economic strength. Such adaptation as occurs would be mainly national and local responses to adverse effects, meaning the large funding requirements of poor and developing countries likely would not be met.

Liberal institutionalism is another important mainstream theory, not as pessimistic as realism and more open to the possibility of system change. Liberal institutionalism, like realism, begins analysis by positing an international system dominated by states. However, the theory sees other significant actors, such as multinational corporations, nongovernmental organizations, intergovernmental organizations, and social movements. Moreover, liberal institutionalism suggests that states can and do pursue the national interest through cooperative arrangements of various kinds. Importantly, liberal institutionalism is open to the idea that the international system changes, so the play of power and the threat of war is not the eternal destiny of humanity.

Much of liberal institutionalist research has focused on "international regimes"[115] as the mechanism for overcoming barriers to cooperation and stabilizing collective efforts to cope with common problems. International regimes institutionalize norms, rules, principles, and procedures that facilitate cooperation in a given issue area. States establish them because they are rational actors and they see cooperation as the rational way to achieve their interests. Thus, liberal institutionalism is more optimistic than realism because it has a different view of rational behavior, not because it expects altruistic behavior by states. In this framework, the climate challenge is an issue area in which an international regime can enable cooperative efforts to mitigate and adapt.[116] Whether institutions do in fact come into being, how they change, and whether they are effective are empirical questions and, regarding climate change, evidence of international institutions making a difference is scarce.

Constructivism emphasizes cognitive factors in the making of world politics. The premise is that human beings form social relations and produce meaning collectively, and failing to take account of that aspect of social life hinders understanding of how international politics are constituted and changed.[117] For constructivism, the material factors stressed in other theories are significant only in regard to how they are defined, understood, and evaluated in social interactions. For example, consider the end of the Cold War. In 1987, the two superpowers held their nuclear arsenals on high alert and relied on mutual assured destruction to deter the other from attacking. This seemed to represent a prime example of how power resources determine relations between states. Yet in a few years, the nuclear antagonists had completely redefined their understandings of one another and had taken their forces off high alert. Nothing in the material circumstances had changed; the change was entirely in the realm of identity and meaning. The United States and the Commonwealth of Independent States that succeeded the Soviet Union came to see one another as friendly rivals rather than bitter enemies, which in turn altered understandings of the importance of material power resources.

Constructivism challenges both realism and liberal institutionalism and their shared assumption of rational actors pursuing their interests. Interests are not given by material factors; they are the product of definitions of self and other, normative consensus and conflicts, and common understandings of appropriate behavior in a given context. Consequently, the critical task for IR theory is to explain and account for interests, not to assume them, and this requires attention to the social construction of reality. For constructivism, revealing the cognitive foundations for identity and interest does most of the work, and analysis of rational action by self-interested states and other actors becomes a rather uninteresting afterthought. Indeed, a more interesting question is how norms become so strong that they constitute background conditions for appropriate social action, taken for granted rather than negotiated anew as issues arise.

Constructivism has addressed various aspects of the climate issue. Examining the character of scientific knowledge indicates that the gap between uncertain scientific understanding of climate change confronts public expectations that science can provide clear, objective, reliable, and authoritative knowledge about important questions.[118] Further, constructivist analysis investigates the manner in which "post-truth" ideas such as climate denialism emerge.[119] In other work, constructivism informs understanding of institutional fragmentation as "normative contestation for strategic social construction."[120] Specifically, the Asia-Pacific Partnership on Clean Development and Climate arose to contest the normative dominance of the UNFCCC with new competing norms, discourses, policies, and ideas. Another influential line of research explores the role of "epistemic communities" in the development of environmental policy and in fostering international cooperation.[121] An epistemic community exists when a body of informed experts reaches a consensus on the nature of a problem and on what should be done about it. The epistemic community exerts influence on policymakers so states arrive at a consensus on which they can agree to take collective action. Presumably, the epistemic community of climate scientists and other experts would provide the knowledge about the causes and effects of climate change, and the appropriate measures to cope with it.

Finally, the diverse, far-ranging body of research going under the broad heading of critical theory, including world-systems theory, Marxism, feminism, and others, investigates the ideological and material aspects of global capitalism that produce ecological crises. These perspectives assert that most theoretical orientations to world affairs reinforce the status quo by not questioning the social relations giving rise to inequality, poverty, racism, sexism, and other injustices. Mainstream research tends to take the world as it is and to conduct analysis of the behavior of individuals, firms, states, and other actors using such tools as statistics to discern patterns, as well as cause and effect. By contrast, critical theory asks how these actors and social relations came into existence, it questions unjust aspects of the existing social order, and it posits possibilities for change and transformation.[122] Envisioning alternative historical trajectories and potential futures is not feasible within the confines of empirical social science reliant on observation of the existing state of affairs. Yet, argue critical theorists, discerning the hidden motors of history, projecting where the world is going, and suggesting where it could go is an essential part of social theory, including research on the social, economic, political, and cultural aspects of climate change. As Matthew Paterson observes, alternative theories "allow us to talk about capitalism, which helps us to place the politics of global warming in a context other than the abstracted formal anarchy of realists and liberal institutionalists."[123]

Ecological Theories

Some students of global environmental politics have argued that the environment presents unique theoretical challenges that cannot be encompassed in international relations theories aimed at explaining all issues in terms of the attributes of the state system. Thus, green theory has emerged, often drawing on other theoretical currents while maintaining the distinct character of environmental politics. Green theory suggests that the environment is more than an issue area; it is an essential prior condition for human society to function, and as such it must be part of the theoretical constructs for comprehending environmental politics. "Marx's account of capitalist production only makes sense when we start to fill in its background conditions of possibility," one of which is a natural world capable of providing natural resources and livable conditions.[124] Green IR theory challenges the mainstream theoretical view that sees the environment as another issue area like all others in four ways. Green IR theory questions the implicit normative content of mainstream theory, casting doubt on its claims to objectivity. In addition, green IR theory observes that rationalist approaches such as liberal institutionalism have not been very successful at explaining global environmental politics. Green IR theory also reveals the actors and institutions that stand in the way of ecological policies, in particular industry-dominated international regimes. Finally, challenging the state-centric orientation of rationalist theory, green IR theory documents the important roles of nonstate actors in environmental norms, policy, and action.

Green theory deploys a number of important concepts to understand global environmental politics. The tragedy of the commons portrays the contradiction between individual incentives and desirable collective outcomes. The tragedy of the commons draws on rational choice models to explain why environmental degradation and overuse of resources occur. The explanation implies, in turn, policy prescriptions. The *commons* refers to any collectively held resource to which access is not limited and overuse can occur. A pasture land, for instance, that is held in common would be available to all farmers wanting to allow their herds to graze. But self-interested maximization of the use of the pasture can result in degradation until no grass remains. Likewise, fishing on the high seas and various other resource issues follow the same logic. The capacity of the planet's atmosphere to absorb GHG is also a common asset, which we are overexploiting, resulting in rising temperatures. The policy responses include such measures as establishing authoritative management of the commons (perhaps even a government), creating individual property rights, and strengthening norms of appropriate behavior.

The concept of externalities points to another problem especially pertinent to environmental issues. In some cases of environmental harms, the problem is that costs are imposed on third parties. A factory produces goods that customers buy, and ideally, the costs of production would be fully reflected in the price of the commodity. But if the factory can dump pollutants on others at little or no cost, it can cut its prices so as to beat the competition, and the price will not reflect all the costs of production. The factory owner and the customer benefit at the expense of those bearing the cost of the pollution. Many environmental issues entail externalities such as transborder air pollution. Adaptation to climate change displays the characteristics of externalities, in that the parties primarily responsible for causing the problem can transfer the costs to those enduring the greatest impacts, those who usually contributed little to causing the problem.

Environmental issues have raised the question of whether economic growth can continue endlessly or, instead, the global economy faces limits to growth. Capitalism, which requires profits to function, appears to need constant economic growth to avoid a crisis. Yet resource and environmental limits might mean that growth cannot continue without end. If growth depends on resource use, then running out of important resources, such as oil, could halt economic growth. By the same token, climate change could create such disruption for human society that normal business cannot carry on as usual.

Beyond limits to economic growth, scientists have asked whether human activities will transcend planetary boundaries, possibly threatening human civilization or even the survival of the species. Of nine critical planetary boundaries, several, including climate change and the loss of biodiversity, have been crossed. Failing to reverse the trends toward crossing more planetary boundaries will be dangerous.

The concept of sustainable development seeks a way to meet these challenges. Sustainable development calls for development that serves the needs of current generations while protecting the rights of future generations. The goal is to conserve resources and rely on renewables so that future generations will have what they need to enjoy a dignified life.

By the 1990s, several forms of green theory had developed. One strand of green theory mounted a critique of capitalism and Soviet-style communism. Both, greens argued, were committed to an ecologically harmful policy of industrialization and growth. Greens rejected the idea that nature was a storehouse of resources for human use, and that nature should be manipulated to provide for human needs, a failing they saw in industrial society, whether capitalist or socialist. Another strand of green theory, drawing on environmental ethics and philosophy, questioned the devaluing of nonhuman creatures and ecosystems; ecocentric and biocentric theory asserts that humans have obligations to treat nature properly.[125] Critical theorists have also addressed the instrumental reason that drives technological industrialism. *Instrumental reason* seeks technical solutions to largely unexamined social ends such as more production. The critical perspective, by contrast, questions whether the ends sought are truly rational in regard to human needs and values. An obsession with growth and development that leads to ecological crisis and possible collapse of human civilization is obviously not rational.[126]

But the fact that so many people are still in need poses a dilemma for green theory. If ecological protection requires zero growth, then how are the pressing needs of those in extreme poverty to be met? Is not economic growth the most feasible answer to poverty today? The green response is that more democracy might improve distributive justice while also curbing ecologically destructive economic growth.[127] Thus, some green theorists have emphasized a more open democratic process by which social ends are chosen and pursued, and this point of view has been incorporated in the IPCC's reports. Further, green theory has increasingly focused on the transnational nature of ecological challenges. This research has produced new conceptions of environmental justice, rights, democracy, activism, citizenship, and governance. Green theory has emphasized the need for environmental justice, which requires inclusiveness, broad participation, application of the precautionary principle, fairness in distributing risks, and compensation for those suffering environmental harms.[128]

Green theory from the left, here labeled ecosocialism, encompasses a wide range of perspectives on how best to address the ecological challenge. As noted, liberal institutionalism considers environmental problems to be issues requiring international cooperation among states, and it often calls for more democracy and appropriate regulation of economic activity. Such views can entail significant reforms, but they do not challenge the fundamentals of the global political economy. Ecosocialism, by contrast, asserts that the ecological crisis cannot be met within existing institutions, that transformation of the global system is necessary to head off ecological collapse. "Capitalism," writes Chris Williams, "is an economic system profoundly and irrevocably at odds with a sustainable planet as it requires ever-greater material and energy throughput to keep expanding."[129]

Several major propositions define the ecosocialist orientation. First, ecosocialists offer a critique of capitalism, intending to show that capitalism creates the ecological crisis and is inherently incapable of solving it. Second, Marxist ecosocialists recover important statements in Karl Marx's work, and themselves advance the development of Marxist thought, to show that Marx and Marxism encompass ecological concerns. Third, ecosocialists argue that the choice before the world is stark: either make a planned transition to a sustainable socialist society, or face the collapse of civilization due to ecological crisis.

Ecosocialists tend to agree that the world has entered a new geological epoch known as the Anthropocene.[130] From the end of the last Ice Age until recently, the world had been in the Holocene, a time with a relatively mild and stable climate that has facilitated the development of agriculture and the rise of civilization. Now, scientists suggest, human impact on the environment has become so great that the planet has entered a new geological era. Humans have changed the climate and changed the global environment in other ways that can be detected just as could previous changes from one period of natural history to the next. The question is whether the Anthropocene will be as compatible with human civilization and survival as the Holocene. Ecosocialists have serious doubts, especially if capitalism remains the prevailing socioeconomic system. Global capitalism will exacerbate the trends that have brought the world to the Anthropocene and cannot solve the problems that will arise in this new world.

Ecosocialist thought highlights that modern capitalist society rests on a consensus on the growth imperative. Capital, of course, demands the conditions for capital accumulation. The state serves this need. Labor, seeking higher wages and more jobs, agrees with capital that the avenue for achieving those ends is economic growth, and it joins capital in demanding that the state maintain pro-growth policies. Thus, the main players in a capitalist society prioritize growth over environment, and support for environmentally sound policies is weak. However, the development of capitalist production and accumulation inevitably begins to reduce working class incomes and impose ever greater environmental costs on communities. As Peter Newell observes, "Perhaps more than any other environmental issue, climate change has forced centre stage the question of the compatibility of capitalism's structural requirement to grow with efforts to keep global warming within tolerable limits."[131] Thus, while expanding production sustains capitalism for the short term, in the long term it creates crisis conditions.

As noted, the first wave of green theory criticized Marxism as well as liberal theory for promoting maximum production of goods and services as the path to a good society. Critics found in Marx's own writings justification for the idea that Marxism celebrated socialism because of its superiority over capitalism in productivity, even though production would be for working people rather than for capitalists. Consequently, critics said, Marxist thought was as guilty of promoting unrestrained exploitation of nature as theories supporting capitalism. Prominent Marxists responded to this critique. They discerned in the work of Marx and other Marxist authors—notably, Marx's lifelong collaborator, Friedrich Engels—resources for a Marxist ecology. John Bellamy Foster and Brett Clark have written at length about Marx's work on the metabolic rift that capitalism creates between society and nature.[132] They cite pithy remarks that Marx and Engels made about capitalism's propensity to degrade the soil and heedlessly exploit nature to accumulate capital. For example, Engels wrote, "Let us not, however, flatter ourselves overmuch on account of our human victories over nature. For each victory nature takes its revenge on us."[133]

Yet these mainly Western ecosocialist Marxists had to address the reality of societies that claimed to be based on Marxist principles, such as the Soviet Union, that also had dismal environmental records. The Chernobyl nuclear accident lifted the veil on environmental damage in the Soviet Union. Marxists asserted that this sorry record did not reflect Marx's views, rather the distortion of Marxism initiated by Joseph Stalin. But the negative example of "actually existing socialism" compiling such a poor environmental record was hard to deny. China's rush toward economic expansion while claiming to be socialist has presented further problems for advocates of ecosocialism.

Even so, ecosocialism insists that capitalism cannot resolve the ecological crisis, at least not in a humane, rational manner. The entire flawed system must be replaced. This will probably require a complete revolution in the hitherto existing mode of production, and simultaneously a revolution in the whole contemporary order, excising the profit motive from society in favor of economic relations that serve the vast majority of workers whose labor is essential to production. Corporate ownership of land must be abolished, and a conscious plan to utilize science and technology in harmony with nature must be implemented.

Despite the poor environmental records of ostensibly socialist societies, the coming ecological crisis induced by capitalism will require an ecosocialist alternative if civilization is to survive.

Advocates claim the society built on ecosocialist principles that would replace capitalism would be truly sustainable. It would embrace renewable energy, efficient use of nonrenewable resources, waste reduction, and production for rational needs rather than artificially created wants. It would end the diversion of material resources to immense, mutually destructive defense spending. A sustainable ecosocialist society would be democratic—debating, adopting, and implementing measures chosen by the majority rather than a handful of corporate, military, and political elites. This would all certainly constitute transformational change. The question of feasibility, however, becomes central to such a program.

Plan of the Book

The central concern of this book is the institutional requisites for effective, equitable adaptation to climate change. The premise for the work is that efforts to reduce global warming have largely failed so far and are likely to see little more success in the future. Consequently, the burden of responding to climate change will fall increasingly on the adaptation side of the ledger. This has become more fully recognized in policy negotiations as well as scholarly literature. While adaptation at one time might have been seen as a less demanding response to global warming than mitigation, it has become clear that adaptation could require profound transformation of global social, political, and economic institutions. However, this transformative potential of adaptation has often been obscured in official discourse.

The approach taken in this book is to recount the development of policy and politics concerning adaptation, to show the points at which adaptation has been tamed and put in service to the institutional status quo, whereas its transformative implications might have been, and still could be, put front and center. Chapter 2 examines the early years of adaptation policy from the landmark 1992 Rio conference through the adoption of the Cancun Adaptation Framework in 2010. Chapter 3 examines the years from Cancun through the important 2015 Paris conference to the present. Chapter 4 considers in depth various understandings of transformation, focusing on institutional change. Chapter 5 examines the relationship between adaptation and capitalism, finding that proposals for some form of ecological modernization are unlikely to meet the challenge. Chapter 6 surveys and analyzes proposals for social transformation, such as ecosocialism, that seek to preserve civilization while putting it on a new, more ecologically sound foundation. The seventh and final chapter considers whether the climate crisis portends an end to civilization and what adaptation means in such a context.

Notes

1. Katharine Hayhoe, *Saving Us: A Climate Scientist's Call for Hope and Healing in a Divided World* (New York: One Signal, 2021), pp. 40–48.
2. IPCC (Intergovernmental Panel on Climate Change), "Summary for Policymakers," in *Climate Change 2023: Synthesis Report. A Report of the Intergovernmental Panel on Climate Change. Contribution of Working Groups I, II and III to the Sixth Assessment Report of the Intergovernmental Panel on Climate Change* (Geneva: IPCC, 2023), p. 4.
3. UNFCCC (UN Framework Convention on Climate Change), *Technical Dialogue of the First Global Stocktake,* FCCC/SB/2023/9, Advance Version, September 8, 2023, pp. 13–14.
4. Steve Vanderheiden, "Introduction," in *Political Theory and Global Climate Change,* edited by Steve Vanderheiden (Cambridge: MIT Press, 2012), p. xi.
5. Fiona Harvey and Damian Carrington, "World Is on 'Highway to Climate Hell,' UN Chief Warns at COP27 Summit," *The Guardian,* November 7, 2022.
6. For an assessment of James Hansen's testimony before Congress in 1988, see Peter Sinclair, "Judgment on Hansen's '88 Climate Testimony: 'He Was Right,'" *Yale Climate Connections,* June 20, 2018.
7. David Spratt and Ian Dunlop, *What Lies Beneath: The Underestimation of Existential Climate Risk* (Melbourne, Australia: National Centre for Climate Restoration, 2018), pp. 10–13.
8. WRI (World Resources Institute), in collaboration with UNDP (UN Development Programme), UNEP (UN Environment Programme), and the World Bank, *World Resources 2010–2011: Decision Making in a Changing Climate—Adaptation Challenges and Choices* (Washington, DC: WRI, 2011).
9. David Schlosberg, "Political Challenges of the Climate-Changed Society," *PS: Political Science and Politics* 46, no. 1 (2013): 13–17.
10. Palitha Kohona, "Climate Change—Are We Really Confronting This Challenge?" *Environmental Policy and Law* 46, no. 2 (2016): 109–111.
11. Paul G. Harris and J. Barkdull, "Adapting to Dangerous Climate Change: Implications for Studies of Politics, Policy and Beyond," *Journal of Environmental Studies and Sciences* 6, no. 1 (2016): 451–459.
12. Katharine Hayhoe, *Saving Us: A Climate Scientist's Call for Hope and Healing in a Divided World* (New York: One Signal, 2021), p. 115.
13. Toby Ord, *The Precipice: Existential Risk and the Future of Humanity* (New York: Hachette Books, 2020), pp. 110–113.
14. Dale Jamieson, *Reason in a Dark Time: Why the Struggle Against Climate Change Failed—and What It Means for Our Future* (New York: Oxford University Press, 2014).

15. IPCC, "Summary for Policymakers," in *Climate Change 2022: Impacts, Adaptation and Vulnerability. Contribution of Working Group II to the Sixth Assessment Report of the Intergovernmental Panel on Climate Change* (Cambridge: Cambridge University Press, 2022), pp. 3–33.

16. Jem Bendell, "Deep Adaptation: A Map for Navigating Climate Tragedy," Occasional Paper No. 2 (Carlisle, UK: Institute for Leadership and Sustainability, 2020), https://www.lifeworth.com/deepadaptation.pdf.

17. Bill McKibben, *Eaarth: Making Life on a Tough New Planet* (New York: Henry Holt, 2010).

18. Karen L. O'Brien and Elin Selboe, eds., *The Adaptive Challenge of Climate Change* (New York: Cambridge University Press, 2015).

19. The social sciences have not been well integrated in climate change research, and political scientists in particular have given scant attention to adaptation to climate change. See Eve Croeser, *Ecosocialism and Climate Justice: An Ecological Neo-Gramscian Analysis* (Abingdon: Routledge, 2021), pp. 17–18; John Barkdull and Paul G. Harris, "Political Science and Severe Climate Change: A Case for Transformational Research into Adaptation," *St. Antony's International Review* 13, no. 1 (2017): 120–140; Debra Javeline, "The Most Important Topic Political Scientists Are Not Studying: Adapting to Climate Change," *Perspectives on Politics* 12, no. 2 (2014): 420–434.

20. IPCC, "Summary for Policymakers," in *Climate Change 2021: The Physical Science Basis. Contribution of Working Group I to the Sixth Assessment Report of the Intergovernmental Panel on Climate Change* (Cambridge: Cambridge University Press, 2021), p. 4.

21. Val Ellicott, "Climate Change: Will World Governments Act in Time?" *CQ Researcher,* October 25, 2021.

22. Katharine Hayhoe, *Saving Us: A Climate Scientist's Call for Hope and Healing in a Divided World* (New York: One Signal, 2021), pp. 182–184.

23. Kevin Anderson, John F. Broderick, and Isak Stoddard, "A Factor of Two: How the Mitigation Plans of 'Climate Progressive' Nations Fall Far Short of Paris-Compliant Pathways," *Climate Policy* 20, no. 10 (2020): 1290–1304.

24. Shaun A. Marcott, Jeremy D. Shakun, Peter U. Clark, and Alan C. Mix, "A Reconstruction of Regional and Global Temperature for the Past 11,300 Years," *Science* 339, no. 6124 (2013): 1198–1201.

25. Global average temperature from the 1880s through the 1910s was about 13.7°C, and was about 14.8°C in 2022. The internationally agreed level at which climate change would present serious danger to human societies is a 2°C increase, or 15.7°C, for a global average temperature nearly 15 percent higher than the period 1880–1919.

26. CO2-Earth, "Annual GHG Index," https://www.co2.earth.

27. José A. Tapia Granados, "Inexorable March Toward Utter Climate Disaster?" *Climate Nature Socialism* 29, no. 4 (2018), p. 25.

28. UN News, "2021 Joins Top 7 Warmest Years on Record: WMO," January 19, 2021.

29. Henry Fountain, "2020 Ties 2016 as Hottest Yet, European Analysis Shows," *New York Times,* January 8, 2021.

30. WMO (World Meteorological Organization), "WMO Confirms that 2023 Smashes Global Temperature Record," January 12, 2024.

31. Fiona Harvey, "World Headed for Irreversible Climate Change in Five Years, IEA Warns," *The Guardian,* November 11, 2011.

32. International Energy Agency, "Oil Market Report—August 2023." https://www.iea.org/reports/oil-market-report-august-2023.

33. Henrik Enroth, "Declarations of Dependence: On the Constitution of the Anthropocene," *Theory, Culture and Society* 38, nos. 7–8 (2020): 1–22.

34. Hannah Ritchie, Max Roser, and Pablo Rosado, "CO2 and Greenhouse Gas Emissions," *Our World in Data,* 2020, https://ourworldindata.org.

35. Sophie Boehm, Katie Lebling, Kelly Levin, Hanna Fekete, Joel Jaeger, Anna Nilsson, Ryan Wilson, Andreas Geiges, and Clea Schumer, *State of Climate Action 2021: Systems Transformations Required to Limit Global Warming to 1.5°C* (Washington, DC: World Resources Institute, 2021).

36. The International Energy Agency reported that demand for coal, the most polluting of the major fossil fuels, would reach record levels, surpassing 8 billion tons for the first time in 2022. In 2000, eight years after the negotiation of the UNFCCC, coal demand stood at about 4.2 billion tons. Thus, despite pledges to abate emissions, the worst offender's output has doubled since nations agreed to address the problem. International Energy Agency, *Coal 2022: Analysis and Forecast to 2025,* December 2022, p. 11, ttps://iea.blob.core.windows.net/.

37. Morgan Bazilian and Dolf Gielen, "5 Years After Paris: How Countries' Climate Policies Match Up to Their Promises, and Who's Aiming for Net Zero Emissions," *The Conversation* December 10, 2020.

38. James Hansen, Pushker Kharecha, Makiko Sato, Valerie Masson-Delmotte, Frank Ackerman, David J. Beerling, and Paul J. Hearty et al., "Assessing 'Dangerous Climate Change': Required Reduction of Carbon Emissions to Protect Young People, Future Generations, and Nature," *PLOS ONE* 8, no. 12 (2013): e81648.

39. Met Office Hadley Centre, "Natural Climate Variability," https://www.metoffice.gov.uk/weather/climate-change/natural-climate-variability.

40. Katharine Hayhoe, *Saving Us: A Climate Scientist's Call for Hope and Healing in a Divided World* (New York: One Signal, 2021), pp. 40–48.

41. Mark New, Diana Liverman, Heike Schroder, and Kevin Anderson, "Introduction: Four Degrees and Beyond: The Potential for a Global Temperature Increase of Four Degrees and Its Implications," *Philosophical Transactions: Mathematical, Physical and Engineering Sciences* 369, no. 1934 (2011): 6–19.

42. Kevin Anderson, "The Hidden Agenda: How Veiled Techno-Utopias Shore Up the Paris Agreement," *kevinanderson.info Blog,* January 6, 2016, kevinanderson.info/blog.

43. Morgan Bazilian and Dolf Gielen, "5 Years After Paris: How Countries' Climate Policies Match Up to Their Promises, and Who's Aiming for Net Zero Emissions," *The Conversation,* December 10, 2020.

44. IPCC, "Long-Term Climate Change: Projections, Commitments and Irreversibility," in *Climate Change 2013: The Physical Science Basis. Contribution of Working Group I to the Fifth Assessment Report of the Intergovernmental Panel on Climate Change* (Cambridge: Cambridge University Press, 2013), pp. 1029–1136.

45. See IPCC, "Long-Term Climate Change: Projections, Commitments and Irreversibility," in *Climate Change 2013: The Physical Science Basis. Contribution of Working Group I to the Fifth Assessment Report of the Intergovernmental Panel on Climate Change* (Cambridge: Cambridge University Press, 2013), fig. 12.5, p. 1054.

46. Rebecca Lindsey and Luann Dahlmann, "Climate Change: Ocean Heat Content," NOAA Climate.gov, August 17, 2020.

47. Paul G. Harris, "COP26: The Eternally Weak Pulse of Climate Diplomacy, and What Needs to Change," *PLOS Climate* 1, no. 3 (2022): e0000019.

48. WMO, "Global Temperatures Set to Reach New Records in Next Five Years," May 17, 2023, https://public.wmo.int. According to the WMO, "There is a 66% likelihood that the annual average near-surface global temperature between 2023 and 2027 will be more than 1.5C above pre-industrial levels for at least one year. There is a 98% likelihood that at least one of the next five years, and the five-year period as a whole, will be the warmest on record."

49. Eamon Akil Farhat, "World Extends Run of Heat Records for an 11th Month in a Row," PhysOrg, May 8, 2024.

50. IISD (International Institute for Sustainable Development), "Secretary-General's SDG Progress Report Proposes Actions to 'Rescue' SDGs," SDG Knowledge Hub, May 25, 2022, http://sdg.iisd.org.

51. Editors, "Notes from the Editors," *Monthly Review* 74, no. 2 (2022), p. 1.

52. IPCC, "Summary for Policymakers," in *Climate Change 2022: Impacts, Adaptation and Vulnerability. Contribution of Working Group II to the Sixth Assessment Report of the Intergovernmental Panel on Climate Change* (Cambridge: Cambridge University Press, 2022), pp. 3–33.

53. IPCC, "Summary for Policymakers," in *Climate Change 2022: Impacts, Adaptation and Vulnerability. Contribution of Working Group II to the Sixth Assessment Report of the Intergovernmental Panel on Climate Change* (Cambridge: Cambridge University Press, 2022), p. 9.

54. IPCC, "Summary for Policymakers," in *Climate Change 2022: Impacts, Adaptation and Vulnerability. Contribution of Working Group II to the Sixth Assessment Report of the Intergovernmental Panel on Climate Change* (Cambridge: Cambridge University Press, 2022), fig. SPM.2, pp. 10–11.

55. IPCC, "Summary for Policymakers," in *Climate Change 2022: Impacts, Adaptation and Vulnerability. Contribution of Working Group II to the Sixth Assessment Report of the Intergovernmental Panel on Climate Change* (Cambridge: Cambridge University Press, 2022), p. 12.

56. IPCC, "Summary for Policymakers," in *Climate Change 2022: Impacts, Adaptation and Vulnerability. Contribution of Working Group II to the Sixth Assessment Report of the Intergovernmental Panel on Climate Change* (Cambridge: Cambridge University Press, 2022), p. 13.

57. IPCC, "Summary for Policymakers," in *Climate Change 2022: Impacts, Adaptation and Vulnerability. Contribution of Working Group II to the Sixth Assessment Report of the Intergovernmental Panel on Climate Change* (Cambridge: Cambridge University Press, 2022), p. 14.

58. IPCC, "Summary for Policymakers," in *Climate Change 2022: Impacts, Adaptation and Vulnerability. Contribution of Working Group II to the Sixth Assessment Report of the Intergovernmental Panel on Climate Change* (Cambridge: Cambridge University Press, 2022), p. 14.

59. IPCC, "Summary for Policymakers," in *Climate Change 2022: Impacts, Adaptation and Vulnerability. Contribution of Working Group II to the Sixth Assessment Report of the Intergovernmental Panel on Climate Change* (Cambridge: Cambridge University Press, 2022), p. 18.

60. Laurie Laybourn, Henry Throw, and Suzannah Sherman, *1.5°C—Dead or Alive: The Risks to Transformational Change from Reaching and Breaching the Paris Agreement Goal* (London: Institute for Public Policy Research and Chatham House, 2023).

61. Mark Lynas, *Our Final Warning: Six Degrees of Climate Emergency* (London: 4th Estate, 2020), p. 167.
62. Committee on Understanding and Monitoring Abrupt Climate Change and Its Impacts, *Abrupt Impacts of Climate Change: Anticipating Surprises* (Washington, DC: National Academies Press, 2013).
63. Jesse Ribot, "Vulnerability Before Adaptation: Toward Transformative Climate Action," *Global Environmental Change* 21, no. 4 (2011): 1160–1162.
64. IPCC, "Emissions Trends and Drivers," in *Climate Change 2022: Mitigation of Climate Change. Working Group III Contribution to the Sixth Assessment Report of the Intergovernmental Panel on Climate Change* (Geneva: IPCC, 2022), p. 237.
65. Chris A. Boulton, Timothy M. Lenton, and Niklas Boers, "Pronounced Loss of Amazon Rainforest Resilience Since the Early 2000s," *Nature Climate Change* 12, no. 3 (2022): 271–278.
66. Jen Flatt Osborn, "Forests Under Threat: A Comprehensive Look at the Latest Deforestation Statistics," World Animal Foundation, April 7, 2023, https://worldanimalfoundation.org.
67. Evan Halper, "Can Vacuums Slow Global Warming? Administration Bets $1.2 Billion on It," *Washington Post,* August 11, 2023; Heinrich Boll Foundation and ETC Group, "Geoengineering Map," 2023, https://map.geoengineeringmonitor.org.
68. Oliver Morton, "The Geopolitical Challenges of Geoengineering—and Geoengineering's Challenge to Geopolitics" (Washington, DC: Wilson Center, September 30, 2020).
69. John Bellamy Foster, "Making War on the Planet: Geoengineering and Capitalism's Creative Destruction of the Earth," *Monthly Review*, 70, no. 4 (2018): 1–10.
70. Paul McLaughlin, "Climate Change, Adaptation, and Vulnerability: Reconceptualizing Societal-Environment Interaction Within a Socially Constructed Adaptive Landscape," *Organization and Environment* 24, no. 3 (2011): 269–291.
71. W. Neil Adger, "Institutional Adaptation to Environmental Risk Under the Transition in Vietnam," *Annals of the Association of American Geographers* 90, no. 4 (2000): 738–758.
72. For example, see Sara Ahmed and Elizabeth Fajber, "Engendering Adaptation to Climate Variability in Gujarat, India," *Gender and Development* 17, no. 1 (2009): 33–50.
73. Adapted from Bonizella Biagini, Rosina Bierbaum, Missy Stults, Saliha Dobardzic, and Shannon M. McNeeley, "A Typology of Adaptation Actions: A Global Look at Climate Adaptation Actions Financed Through the Global Environment Facility," *Global Environmental Change* 25 (2014): 97–108; UNEP, *Adaptation Gap Report 2022: Too Little, Too Slow—Climate Adaptation Failure Puts World at Risk* (Nairobi: UNEP, 2022);USGCRP (US Global Change Research Program), "Reducing Risks Through Adaptation Action," in *Impacts, Risks, and Adaptation in the United States: Fourth National Climate Assessment,* vol. 2 (Washington, DC: USGCRP, 2018): 1309–1345, DOI 10.7930/NCA4.2018.
74. IPCC, "Summary for Policymakers," in *Climate Change 2022: Impacts, Adaptation and Vulnerability. Contribution of Working Group II to the Sixth Assessment Report of the Intergovernmental Panel on Climate Change* (Cambridge: Cambridge University Press, 2022), pp. 5–6.
75. Benjamin K. Sovacool and Björn-Ola Linnér, *The Political Economy of Climate Change Adaptation* (New York: Palgrave Macmillan, 2016), p. 7.
76. IPCC, "Climate-Resilient Pathways: Adaptation, Mitigation, and Sustainable Development," in *Climate Change 2014: Impacts, Adaptation, and Vulnerability.*

Part A: Global and Sectoral Aspects. Contribution of Working Group II to the Fifth Assessment Report of the Intergovernmental Panel on Climate Change (Cambridge: Cambridge University Press, 2014), p. 1122.

77. Billie L. Turner II, "Vulnerability and Resilience: Coalescing or Paralleling Approaches for Sustainability Science," *Global Environmental Change* 20, no. 4 (2010): 570–576.

78. IPCC, "Climate-Resilient Pathways: Adaptation, Mitigation, and Sustainable Development," in *Climate Change 2014: Impacts, Adaptation, and Vulnerability. Part A: Global and Sectoral Aspects. Contribution of Working Group II to the Fifth Assessment Report of the Intergovernmental Panel on Climate Change* (Cambridge: Cambridge University Press, 2014), p. 1122.

79. Eriksen, Siri H., Andrea J. Nightingale and Hallie Eakin, "Reframing Adaptation: the political nature of climate change adaptation," *Global Environmental Change* 35 (2015), p. 531.

80. The author saw variable effects of exposure personally in the summer of 2023. After living in Texas for thirty years, we moved to Illinois just in time to experience winds as high as 100 miles per hour sweeping across Springfield. Texas is generally ready for high winds, but Springfield was not, and much of the city was without power for days. The upshot now is the city is exploring ways to reduce its vulnerability to high winds, exposure to them having revealed serious vulnerabilities in the city's energy infrastructure. Of course, attributing a single weather event to climate change is not possible, but climate scientists say that the probability of such events has risen with temperatures.

81. John Anderies, Carl Folke, and Elinor Ostrom, "Aligning Key Concepts for Global Change Policy: Robustness, Resilience, and Sustainability," *Ecology and Society* 18, no. 2 (2013), p. 8.

82. IPCC, "Summary for Policymakers," in *Climate Change 2022: Impacts, Adaptation and Vulnerability. Contribution of Working Group II to the Sixth Assessment Report of the Intergovernmental Panel on Climate Change* (Cambridge: Cambridge University Press, 2022), p. 5.

83. Melinda Harm Benson and Robin Kundis Craig, *The End of Sustainability: Resilience and the Future of Environmental Governance in the Anthropocene* (Lawrence: University of Kansas Press, 2017).

84. W. Neil Adger, "Vulnerability," *Global Environmental Change* 16, no. 3 (2006): 268–281.

85. IPCC, "Summary for Policymakers," in *Climate Change 2022: Impacts, Adaptation and Vulnerability. Contribution of Working Group II to the Sixth Assessment Report of the Intergovernmental Panel on Climate Change* (Cambridge: Cambridge University Press, 2022), p. 11.

86. Isabel van de Sand, "Payments for Ecosystem Services in the Context of Adaptation to Climate Change," *Ecology and Society* 17, no. 1 (2012), p. 11.

87. Nick Brooks and W. Neil Adger, "Assessing and Enhancing Adaptive Capacity," in *Adaptation Policy Frameworks for Climate Change: Developing Strategies, Policies and Measures,* edited by Bo Lim and Erika Spanger-Siegfried (Cambridge: Cambridge University Press, 2004), p. 168.

88. Stephen R. Carpenter and William A. Brock, "Adaptive Capacity and Traps," *Ecology and Society* 13, no. 2 (2008), p. 40.

89. Nathan L. Engle, "Adaptive Capacity and Its Assessment," *Global Environmental Change* 21, no. 2 (2011): 647–656.

90. Nick Brooks and W. Neil Adger, "Assessing and Enhancing Adaptive Capacity," in *Adaptation Policy Frameworks for Climate Change: Developing Strategies,*

Policies and Measures, edited by Bo Lim and Erika Spanger-Siegfried (Cambridge: Cambridge University Press, 2004), p. 168.

91. David Harvey, *A Brief History of Neoliberalism* (Oxford: Oxford University Press, 2005); Jack Rasmus, *The Scourge of Neoliberalism: US Economic Policy from Reagan to Trump* (Atlanta: Clarity Press, 2020).

92. Jason W. Moore, *Capitalism in the Web of Life: Ecology and the Accumulation of Capital* (London: Verso, 2015).

93. Edward R. Carr, "Properties and Projects: Reconciling Resilience and Transformation for Adaptation and Development," *World Development* 122 (October 2019): 70–84.

94. IPCC, "Summary for Policymakers," in *Climate Change 2022: Impacts, Adaptation and Vulnerability. Contribution of Working Group II to the Sixth Assessment Report of the Intergovernmental Panel on Climate Change* (Cambridge: Cambridge University Press, 2022), pp. 32–35.

95. IPCC. *Climate Change 2022: Impacts, Adaptation, and Vulnerability. Contribution of Working Group II to the Sixth Assessment Report of the Intergovernmental Panel on Climate Change* (Cambridge University Press, 2022), p. 179.

96. Eriksen, Siri H., Andrea J. Nightingale, and Hallie Eakin. "Reframing Adaptation: The Political Nature of Climate Change Adaptation," *Global Environmental Change* 35 (November 2015), p. 526.

97. For an analysis of the barriers to adaptation policy, see Susanne C. Moser and Julia A. Ekstrom, "A Framework to Diagnose Barriers to Climate Change Adaptation," *Proceedings of the National Academy of Sciences of the United States of America* 107, no. 51 (2010): 22026–22031.

98. IPCC, *Climate Change 2014: Impacts, Adaptation, and Vulnerability. Contribution of Working Group II to the Fifth Assessment Report of the Intergovernmental Panel on Climate Change* (New York: Cambridge University Press), p. 1122.

99. IPCC (2022), *Climate Change 2022: Impacts, Adaptation and Vulnerability: Contribution of Working Group II to the Sixth Assessment Report of the Intergovernmental Panel on Climate Change* (Cambridge: Cambridge University Press), p. 2657.

100. IPCC (2022), *Climate Change 2022: Impacts, Adaptation and Vulnerability: Contribution of Working Group II to the Sixth Assessment Report of the Intergovernmental Panel on Climate Change* (Cambridge: Cambridge University Press), p. 123.

101. IPCC (2022), *Climate Change 2022: Impacts, Adaptation and Vulnerability: Contribution of Working Group II to the Sixth Assessment Report of the Intergovernmental Panel on Climate Change* (Cambridge, UK, and New York: Cambridge University Press), p. 2658.

102. IPCC (2022), *Climate Change 2022: Impacts, Adaptation and Vulnerability: Contribution of Working Group II to the Sixth Assessment Report of the Intergovernmental Panel on Climate Change* (Cambridge, UK, and New York: Cambridge University Press), p. 2659.

103. With rare exceptions, the IPCC avoids even using the term *capitalism*. This aversion is also present in academic research. See Peter Newell, "The Elephant in the Room: Capitalism and Global Environmental Change," *Global Environmental Change* 21, no. 1 (2011): 4–6.

104. Ted Nordhaus, Vijaya Ramachandran, and Patrick Brown, "The Obvious Climate Strategy Nobody Will Talk About," *Foreign Policy*, November 6, 2022.

105. IPCC, *Climate Change 2022: Impacts, Adaptation and Vulnerability. Contribution of Working Group II to the Sixth Assessment Report of the Intergovernmental*

Panel on Climate Change (Cambridge: Cambridge University Press, 2022), pp. 2495–2496.

106. William J. Ripple, Christopher Wolf, Thomas M. Newsome, Phoebe Barnard, and William R. Moomaw, "World Scientists' Warning of a Climate Emergency," *BioScience* 70, no. 1 (2020): 8–12.

107. Johan Rockström, "A Resilient Earth for Future Generations." In World Wide Fund for Nature, *Living Planet Report 2016: Risk and Resilience in a New Era,* edited by World Wide Fund for Nature (Gland: WWF International, 2016).

108. Peter Dauvergne, "The Coming Crisis: Planetary Instability," speri.comment: *The Political Economy Blog,* April 27, 2016, http://speri.dept.shef.ac.uk.

109. Leslie Sklair, "Social Movements for Global Capitalism: The Transnational Capitalist Class in Action," *Review of International Political Economy* 4, no. 3 (1997): 514–538.

110. Leslie Sklair, *The Anthropocene in Global Media: Neutralizing the Risk,* Routledge Studies in Environmental Communication and Media (Abingdon: Routledge, 2021), p. 6.

111. John Barkdull and Paul G. Harris, "Climate Catastrophe and Transformationalism," *Seton Hall Journal of Diplomacy and International Relations* 16, no. 1 (2015): 119–130.

112. Jason W. Moore, *Capitalism in the Web of Life: Ecology and the Accumulation of Capital* (London: Verso, 2015).

113. Matthew Paterson, *Global Warming and Global Politics* (London: Routledge, 1996).

114. Kenneth N. Waltz, *Theory of International Politics* (Reading, MA: Addison-Wesley, 1979).

115. Stephen Krasner, ed. *International Regimes* (Ithaca: Cornell University Press, 1983); Mark W. Zacher and Richard A. Matthew, "Liberal International Theory: Common Threads, Divergent Strands," in *Controversies in International Relations Theory: Realism and the Neoliberal Challenge,* edited by Charles W. Kegley Jr. (New York: St. Martin's Press, 1995), 107–150.

116. Robert O. Keohane and David G. Victor, "The Regime Complex for Climate Change," *Perspectives on Politics* 9, no. 1 (2011): 7–23.

117. Alexander Wendt, *Social Theory of International Politics* (Cambridge: Cambridge University Press, 1999).

118. Philipp Aufenvenne, Heike Egner, and Kirsten von Elverfeldt, "On Climate Change Research, the Crisis of Science and Second-Order Science," *Constructivist Foundations* 10, no. 1 (2014): 120–139.

119. Frank Fischer, "Knowledge Politics and Post-Truth in Climate Denial: On the Social Construction of Alternative Facts," *Critical Policy Studies* 13, no. 2 (2019): 133–152.

120. Shunji Matsuoka, "The Genesis and End of Institutional Fragmentation in Global Governance on Climate Change from a Constructivist Perspective," *International Environmental Agreements: Politics, Law and Economics* 17, no. 2 (2017): 143–159.

121. Peter M. Haas, *Saving the Mediterranean: The Politics of International Environmental Cooperation* (New York: Columbia University Press, 1990).

122. Robert W. Cox, "Social Forces, States and World Orders: Beyond International Relations Theory," *Millennium—Journal of International Studies* 10, no. 2 (1981): 126–155.

123. Matthew Paterson, *Global Warming and Global Politics* (London: Routledge, 1996), p. 7.

124. Nancy Fraser, "Behind Marx's Hidden Abode," *New Left Review* 86 (2014), p. 60.

125. John Barkdull and Paul G. Harris, "The Land Ethic: A New Philosophy for International Relations," *Ethics and International Affairs* 12, no. 1 (1998): 159–177.

126. Arran Gare, *Nihilism Incorporated: European Civilization and Environmental Destruction* (Bungendore, Australia: Eco-Logical Press, 1993).

127. John Barry, "Green Republicanism and a 'Just Transition' from the Tyranny of Economic Growth," *Critical Review of International Social and Political Philosophy* 24, no. 5 (2019): 725–742.

128. Alan R. Johnson, "Avoiding Environmental Catastrophes: Varieties of Principled Precaution," *Ecology and Society* 17, no. 3 (2012), p. 9.

129. Chris Williams, *Ecology and Socialism* (Chicago: Haymarket Books, 2010), p. 175.

130. Ian Angus, *Facing the Anthropocene: Fossil Capitalism and the Crisis of the Earth System,* Kindle ed. (New York: Monthly Review Press, 2016).

131. Peter Newell, "The Elephant in the Room: Capitalism and Global Environmental Change," *Global Environmental Change* 21, no. 1 (2011): 4–6.

132. John Bellamy Foster, Brett Clark, and Richard York, *The Ecological Rift: Capitalism's War on the Earth* (New York: Monthly Review Press, 2010).

133. Frederick Engels, *The Origin of Family, Private Property and the State* (New York: International Publishers, 2007), pp. 260–261.

2

The Turn to Adaptation

THIS CHAPTER DESCRIBES AND ANALYZES THE RISE OF ADAPTation on the global agenda, from the expectation that it would be the default response, to seeing it as a distraction from the main task of reducing emissions, and then to viewing it as an important remedy to the challenge of climate change. Lisa Dale observes that only recently has "the notion of adapting to a changing climate thoughtfully, systematically and with advance planning" become a global priority.[1] Yet the rise of adaptation has been long term and steady, not a sudden policy shift, marked by significant changes in how various actors view adaptation as a response to the climate crisis. In this chapter, important landmarks in the rise of climate adaptation policy are identified and examined. They include:

- The introduction of adaptation to the climate agenda leading up to the 1992 Rio conference and adoption of the UN Framework Convention on Climate Change (UNFCCC);
- Recognition of adaptation as a significant element of climate policy at the Marrakesh 2001 Conference of the Parties (COP)7;
- Adoption of the 2003 Buenos Aires Program of Work on Adaptation and Response Measures at COP10;
- Agreement to the Nairobi Work Programme on Impacts, Vulnerability and Adaptation to Climate Change 2006;
- Negotiation of the 2007 Bali Action Plan adopted at COP13;
- Adoption of the Cancun Adaptation Framework at the 2010 COP16.

The Changing Dialogue on Adaptation

The general outlines of the climate challenge had already come into view well before governments agreed to take measures to reduce greenhouse gases (GHG) emissions. Concerns about human influence on the global climate had emerged as early as the 1960s, received mention in the 1972 Stockholm declaration on the environment,[2] and later led the UN General Assembly to call, in 1974, for the World Meteorological Organization (WMO) to study the matter. The WMO dismissed then current speculation on global cooling and "reaffirmed the general scientific expectation of greenhouse warming."[3]

The First World Climate Conference

The WMO convened the first World Climate Conference in Geneva in 1979, outlining a research agenda and spurring the creation of the World Climate Programme (WCP) and the World Climate Research Programme (WCRP). The conference concluded by asserting the need for greater knowledge about the climate and its effects on human society, and research to foresee and prevent possible human-caused changes in the climate that could harm human interests. The ensuing WCP and WCRP quickly became interdisciplinary and inter-agency efforts, with the UN Environment Programme taking the lead on the study of impacts, and WMO joining the International Council for Science (ICSU) in leading the climate science research.

The participants in the 1979 conference were already employing the terms *vulnerability* and *resilience* to describe the dimensions of the climate challenge.[4] They tended to assume that human society would seek ways to limit the damage, viewing adaptation as a logical and probable element of the response to climate change.[5] For instance, Evgenii K. Fedorov outlined a three-part strategy to enable the world cope with the negative impacts of such climate change as might occur. The first two parts included developing the capacity to predict future climate conditions and to assess the effects on societies. The third element of the strategy would be to seek ways to avoid adverse consequences (adapt), or to prevent climate change from occuring (mitigate). Thus, in Fedorov's view, adaptation was the likely first option, with mitigation taking more or less equal place.

Interestingly, Fedorov also raised the institutional question that now draws the attention of much scholarship and policy analysis on adaptation. He noted that Russia faced severe droughts three times, the first two (1831 and 1921) resulting in widespread suffering, while the Soviet state successfully *increased* food production during a 1972 drought. Because, Fedorov argued, the Soviet Union was a developed socialist country rather than a czarist empire or in the middle of a civil war, similar climatic conditions resulted in very different outcomes. That said, Soviet capacity to cope with the 1972 drought depended in part on access to outside resources, and Soviet purchases were large enough to upset world grain markets. Even so, the attention to institutional questions is remarkable. Further, to achieve desirable adaptation to highly probable new climatic conditions, Fedorov concluded, would require lasting peace and an end to the arms race; that is, major changes to global institutions and practices, not only technocratic, incremental measures.[6]

However, other authors saw adaptation occurring without such radical change. For example, writing about the science of climate change, W. Lawrence Gates noted that human societies have long known how to adjust to varying climatic conditions around the world and to gradual changes in average temperature and rainfall. Moreover, this "adaptation to the climate is so much an accepted part of life" that we are only made aware of our relationship to the climate during crises such as droughts or other extreme weather.[7] Accordingly, while uncertainty remained about the amount and effects of future warming, it would be reasonable to believe that humans would adapt to new climate conditions as they have habitually done. Likewise, Gates called for research into the dimensions of climate variability and "the range of potential human adjustment," not a program to reduce GHG emissions to avert global warming. Indeed, many of the scientists attending the 1979 conference believed that then observed climate extremes were in line with historical variability, thus not requiring major changes to the energy system or institutions. Moreover, according to the conference chair, regardless of the long-term trend toward a warmer world, increased vulnerability to climate variability resulted from modernization and industrialization, as well as the much wider extent of human settlement, trends that would continue regardless of temperature rises. Robert M. White observes that the "perception of broader climatic vulnerability" arises from "world population growth, increased world demand for food, energy, and other resources, increased interdependence of nations, and the pace of economic development." Consequently, "vulnerability can only increase because the underlying causes will intensify, not diminish." This might suggest a contradiction with the current belief that vulnerability results from too little development, not too much. White is suggesting that, regardless of whether temperatures rose, which was an uncertain proposition at the time, more development would mean more vulnerability. To be sure, significant climate change would mean more threats to society in addition to those resulting from more people in more highly developed countries. But the source of rising vulnerability to the entire range of environmental challenges would be the increasing pace of development.[8] The conference called for continued research to establish whether human activities were having an effect on the global climate and what the impacts of climate change would be; the final outcome of the conference affirmed the judgment that increasing GHG would likely result in rising temperatures with significant regional impacts. However, there was little sentiment for strong, immediate measures to reduce the sources of climate change, when human society would likely adapt anyway. Obviously, then, there was no consideration of transformational change in global institutions.

In 1985, the fruits of further research efforts were expressed at the Villach conference, which publicized scientific opinion that the first half of the twenty-first century would see temperature rises greater than any before experienced in human history. The conference, in addition to calling for continued research, suggested the possibility of a global convention on the climate challenge. By this time, consensus was forming around the need for emissions limits. But in much of the report on the conference, it is taken for granted that societies will have to find ways to adjust to a warmer world. For instance, an assessment of agriculture considered how farming would change under different climate scenarios, implicitly assuming that change was coming. To be sure, the conference discussed GHG reductions as an option. Yet it also stated that significant scientific uncertainty remained about the dimensions of the problem. Consequently, Working Group I on Emissions of CO_2 into the Atmosphere, while calling for more research, added that "the implementation of policy options to modify or control the use of fossil fuels is not warranted on the basis of the climate change problem alone."[9] Instead of the emphasis on emissions reductions that has prevailed since, the conference asked whether the global warming likely to come would inflict damage beyond the capacity of human society to adjust. Consequently, the necessity of emissions controls remained an open question.

Certainly, the 1985 Villach conference represented a growing consensus on the probability of a warmer planet in the coming decades and that human activities were major contributors to observed trends. Consequently, the conference was an important spur to increasing policy attention to the matter. Yet the scientific consensus of the time did not include strong recommendations for emissions reductions, nor for fundamental transformation of the global energy system. Such steps were raised as possibilities, but alongside reliance on the capacity of human society to cope with altered climate conditions.

The findings presented at the Villach conference made their way into the Brundtland Report on sustainable development, and they added momentum to creation of the Intergovernmental Panel on Climate Change (IPCC). By 1987 and the convening of the Tenth World Meteorological Congress, participants had agreed that governments, not scientists in their individual capacities, should be providing overall guidance to climate research and assessment. The IPCC was created in 1988 to undertake this role.

Participants in these early discussions were considering adaptation as a likely, logical response to a changing climate. For instance, Irving Mintzer of the World Resources Institute explained the implications of four policy scenarios: continuing on then current tracks, shifting to even higher emissions, adopting incremental policies to slow emissions, and implementing a more aggressive policy track toward significantly slower accumulation of GHG. He concluded, in terms that resonate still, "Clearly, policy choices made today and implemented soon will substantially affect the magnitude and speed of planetary warming."[10] Significantly, Mintzer noted the three options available to policy makers: wait and see, reduce emissions, *or adapt*. Mintzer went on to observe, "The choice is not between preventing or adapting to climate change. The challenge is to find those policies that, in circumstances peculiar to each region and nation, will slow the rate of change and allow societies to adapt to those climatic changes which cannot be avoided."[11] Thus, a prominent figure in environmental affairs had outlined the challenge years before adoption of the UNFCCC, presenting adaptation as one of the major policy options for addressing the growing problem. However, Mintzer and others proposed adaptation within the existing social, economic, and cultural institutional framework. Thus, although adaptation appeared early in scientific and policy discussions, its radical implications were little recognized.

The Second World Climate Conference

The second World Climate Conference took place in Geneva in 1990, soon after the release of the first IPCC Assessment Report (AR1).[12] In the few intervening years, the tone of scientific opinion had changed considerably, toward stronger advocacy of emissions reductions. Noting that the previous statement by leading scientists had called mainly for further research, the 1990 document stated bluntly that states should act immediately to reduce net emissions through national, regional, and global policy coordination with the goal of ensuring that the concentration of greenhouse gases remained below the level creating risks for nature and human society.[13]

Still, although the experts made much stronger calls for reducing emissions, they continued to place stress on adaptation to the effects of a changing climate. While pressing for emissions reductions, the report on the 1990 meeting assumed that humans would have to make significant adjustments. Indeed, preparing for future climate conditions was the main reason given for investing in research that would enable increasingly accurate predictions for planning purposes. Improved forecasting of climate variability, noted the conference report, "would provide enormous economic and welfare benefits in coping with droughts, prolonged rain, and periods of severe hot and cold weather,"[14] plainly assuming that those things would happen and society would have to adapt to them.

The management of water resources and irrigation should be designed, said the report, to account for predicted changes in climatic conditions; that is, anticipatory adaptation was needed. Likewise, land use and urban development would require planning for expected climate conditions, including improved building construction that would enhance both efficiency and resilience in the face of more challenging conditions. For its part, said the experts, agriculture would be affected by population growth and investments in agricultural productivity more than by climate change; meeting future food needs thus meant "there should be an effort to reduce population in developing countries and reduce consumption in developed countries,"[15] recommendations that would apply regardless of climate change. Thus, from the first conference to the second, the emphasis on mitigation rose, but the expectation that adaptation would play an important part in the response to climate change remained. The scientific consensus of the time took for granted that human society would likely undertake adaptation as its main response to the effects of rising global average temperature.

The full report of the 1990 conference devotes a chapter to adaptation measures, observing "considerable optimism among many that adaptation to climate change is feasible, since it is as old as the human species."[16] It refers to the widespread expectation that human society would adapt to climate change, as it had done to natural changes many times before. The chapter notes, moreover, the capacity to adapt had increased due to technological progress and greater prosperity. To be sure, adaptive capacity varies, largely corresponding to a country's level of development, but diverting additional resources to poorer countries could presumably overcome barriers to adaptation. In sum, the tone of this part of the conference report was optimistic regarding adaptation's potential to redress harms from climate change.

In the same vein, the summary of Working Group II's findings asserts that "the calculations and assessments made did not point to any catastrophic socio-economic consequences on *global* scales,"[17] which suggests that no global response was required and the international community could decide whether to provide assistance to those areas hard-hit by climate change. The absence of the threat of global catastrophe suggests that most of the world would adjust, albeit at some cost, to changing conditions, with the costs in some regions being quite high, in others low. In general, the consensus as represented in the first reports of the IPCC and other sources was that serious impacts were likely and humans would react to them as they occurred. It would be wise for urban planners, engineers, and others to recognize that they could not assume conditions would be the same in the future as in the past, but adaptation in some form would be the default response. This is not to say that no alarm bells were going off. To the contrary, James Hansen's widely publicized appearance before Congress warning of serious impacts to come had occurred in 1988, and scientists were calling for mitigation to avert the worst impacts of global warming. Nevertheless, the heavy emphasis on mitigation seen later was not yet present in the scientific consensus, which expected humans' significant adaptation capacity would cope with many of the coming challenges. Indeed,

> anthropogenic climate changes are forecast to occur within the next 40 to 100 years. Such time spans are too long for the people to perceive or to arouse and sustain their attention. Attention will gradually fade in as much as adaptation of the people and adjustments of society and environment are approaching a better fit with climatic conditions. Baseline values will shift and old ones are soon forgotten. Irrespective of the patterns of change there is ample time during periods of one to two human lifespans to adjust tolerance bands and safety margins.[18]

That is, climate change would bring about new conditions incrementally, not abruptly. The pace of change would be extremely rapid on geological scales, but it would be nearly imperceptible from year to year on human scales, and people would adjust to a gradually warming world. This is not an urgent call for transforming the global energy system. To the contrary, it suggests that people and societies will gradually adjust as the climate gradually changes, implying that policy interventions could be minimal.

To be sure, the Conference Statement did urge strong mitigation measures, but many parts of the report, which was the most authoritative statement on the scientific consensus of the time, justified further research because of its contribution to adaptation planning and expected considerable adaptation to occur. As the passage quoted above indicates, the feasibility of adapting to a warmer world was not much in doubt, and warnings about the collapse of civilization were sparse. This changed when political leaders fully confronted the challenge in the negotiation of international agreements governing climate policy.

The George H. W. Bush Administration

US climate policy during this critical period underwent significant change.[19] In addition, the Republican administration's policy statements provide a gauge to the views of the business community during this time. President George H. W. Bush had campaigned on addressing the climate crisis, declaring in a 1988 speech that he would convene a meeting of world leaders at the White House to that end. After taking office, in May 1989 Bush said: "I expect that these efforts will lead to formal negotiations on the establishment of a framework convention on global climate. It is important that this process led to international scientific consensus on the seriousness of the issue for the environment and for the world economy. At the same time, we should ensure that the interests of developing countries are taken into account in this process."

In June 1989, Bush recognized growing concern about "the warming of the world's climate," and his belief that global cooperation could address the range of environmental challenges.

However, the administration emphasized the uncertainty in the science, which meant that the country should delay before committing to costly measures. Thus, the administration called for more research rather than policy action, although it was already in favor of an international convention to address the problem. Indeed, in October 1989 Bush responded to a reporter's question on US leadership by asserting that the United States would continue to lead "based on the finest, most up-to-date science possible," with the country providing "that kind of leadership *in science*" (emphasis added) while avoiding policy extremes.

Yet Bush's chief of staff, former New Hampshire governor John Sununu, strongly opposed limitations on fossil fuel development, and he brought conservatives from think tanks into the conversation to rebut the growing scientific consensus. By 1990, Bush was putting emphasis on preserving jobs and economic growth, insisting that environmental protection and economic prosperity would work together. Bush challenged what he called extremists who "don't want this country to grow," and he claimed they were threatening to throw people out of work. In his April 1991 message to Congress on the environment, Bush called for reliance on free markets to the greatest extent possible. He reiterated support for an international convention on climate change, which, he said, should comprehensively include all sources and sinks of greenhouse gases, as well as adaptation measures, with the usual calls for further research. In March of the 1992 election year, Bush declared he was seeking to "merge economic and environmental goals." He said again that a convention on climate should encompass adaptation measures to be included in national action plans. Bush deplored "carbon-dioxide-only proposals that others have espoused," rejecting "an exclusive focus on targets and timetables." In that context, adaptation would play a greater role in climate policy.

In autumn 1991, the US National Academy of Sciences produced a report asserting that the United States could adapt to most climate change impacts without great inconvenience. With some dissent, the "adaptation panel" found the rate and amount of change likely to occur by the middle of the twenty-first century was well within the adaptive capacity of the country. Observed Leslie Roberts, "The panel, which was chaired by Paul Waggoner of the Connecticut Agricultural Experiment Station, has few doubts that society, at least, can keep up. Indeed, the report reads like a virtual paean to human adaptability, technological innovation, and the market."[20] While acknowledging that natural ecosystems and developing countries would face serious challenges, the panel "found no cause for alarm" in projections of rising average temperatures.[21] Undoubtedly, this prominent study from a reputable source was congenial to the fossil fuel lobby and its friends in government.

Other figures associated with the government were making similar arguments, warranting reliance on adaptation and cautious, no-regrets approaches to mitigation. For example, influential economist William Nordhaus claimed that aggressive mitigation would be unwise before the effects of global warming were known, for the costs of significant emissions cuts could exceed the benefit to be reaped years in the future as the planet warms. Better, he said, to invest in economic growth so adaptation would be well implemented to cope with problems as they arose.[22] Nordhaus observed, "Climate change is likely to produce a combination of gains and losses with no strong presumption of substantial net economic damages. This is not an argument in favor of climate change or a laissez-faire attitude to the greenhouse effect. Rather, it suggests that a careful weighing of costs and damages will be necessary if a sensible strategy is to be devised."[23]

Nordhaus might have said that his work should not justify a laissez-faire attitude, but his finding that doing nothing but letting markets respond would probably cost less than 1 percent of gross domestic product (GDP) at mid-century did encourage just that attitude.[24] Regardless of his intent, the policy message was that the best course was to cross the adaptation bridge when the country came to it, and the best vehicle for crossing the bridge would be a growing, prosperous economy built on fossil fuels.

Rio 1992 to Buenos Aires 2004

Contrary to the George H. W. Bush administration's orientation, the Rio Earth Summit marked the international community's focus on mitigation and relegation of adaptation to a secondary role. The UNFCCC called for strong measures on mitigation so that natural systems would have time to adapt, which would require slowing the pace of temperature rise and capping it at a noncatastrophic level. The growing resistance of the fossil fuel industry to limits on production and use of oil, coal, and gas included claims that adaptation could cope with all but the most extreme climate change. This led climate activists to deemphasize adaptation so as to remove an argument against mitigation. Policy reflected a kind of middle ground, accepting that some adaptation would have to occur, but that it would only be feasible if temperature increases were kept in a safe zone; that is, below 2°C. Consequently, achieving sufficient reductions in GHG emissions to hold the temperature line was the top priority even for the compromise view that policy demanded "mitigation plus adaptation." The problem with this view was the growing evidence that mitigation would fail to achieve its goals. This cast doubt on the feasibility of adaptation and the capacity of prevalent social, political, and economic institutions to avert a crisis. These concerns became more salient in the decade following the Rio conference.

With time, developing countries would prove ready to take up the banner for adaptation. If that was what powerful actors like the United States wanted, developing countries insisted the wealthy nations, the same ones that had created the problem, should pay the bill. Thus, while mitigation took the spotlight, adaptation did not disappear from the scene. But adaptation, unlike earlier discussions among scientists, became a minor theme in official negotiations.

The complexities of the adaptation discourse call for a closer look at the landmark 1992 conference in Rio.

What Happened at Rio 1992?

Called by some the last chance to save the world, hailed by a few as a turning point in global environmental action, deemed by many to have been a failure, the UN Conference on Environment and Development held in Rio de Janeiro saw the negotiation of the first major international agreement on climate change. Yet resistance to strong environmental policy had already formed. Facing a global upsurge of environmental concern and the demands of developing countries that the developed countries pay for remedies, major powers held confidential meetings on how to thwart the ostensible main aims of the conference.[25] Accordingly, the first cut at a climate accord fell well short of enacting sufficient mitigation policy while barely mentioning adaptation.

Even the most hopeful concept embraced at the Rio conference—sustainable development—came under fire, as David Orton exemplifies in his call for "sustainability." Orton argued that sustainable development was a recipe for "global ecological disaster," whereas a proper response to the ecological crisis would require a massive redistribution of wealth, new ethics, and new institutions oriented toward ecological consciousness, democracy, and cultural diversity.[26]

This radical vision of sustainability was not an important part of the Rio negotiations, and the more limited goals for sustainable development found in official negotiations are still far from fulfillment. Regarding the question addressed in this study, the documents emerging from the meeting and the most heavily discussed agenda items show little attention to adaptation to climate change, compared to the attention given to reducing GHG emissions. But that attention has not translated into effective mitigation policy, as recent data show.[27]

The record of negotiations. The Rio Earth Summit addressed a range of environmental challenges, including those that were topics of new international conventions: forests, biodiversity, and climate change. Negotiations took place over the preceding two years in meetings of the Preparatory Committee (PrepCom) and two Intergovernmental Negotiating Committees (INC). One INC addressed biodiversity, and the other worked on the Framework Convention on Climate Change. Four major meetings of the PrepCom produced Agenda 21 and the Rio Declaration, which included sections on the climate challenge. These negotiations left much to be decided, so the Rio Summit became, instead of a "massive photo opportunity . . . a critical negotiation session."[28]

"Of all the issues involved," noted Matthew Paterson and Michael Grubb in advance of the conference, "climate change is widely believed to be the acid test of whether or not countries are serious, because the responses required to limit and manage climatic change could go to the heart of their political and industrial structure."[29] However, as was soon apparent, the agreement reached on climate change barely touched the political and industrial structures of countries or the world. "So here we are," wrote George Monbiot, "after 50 years of engineered failure, with not one of the 40 markers of climate action on track to meet the targets governments have agreed."[30]

Paterson and Grubb explained that significant barriers to agreement revealed in prior negotiations were likely to grow more difficult to surmount as negotiations developed. Barriers they identified in advance of Rio are, to varying degrees, still present today. The barriers include the widely different implications of alternative quantitative measures of a country's contribution to emissions (annual, historical, per capita); limits on the technical ability to reduce emissions; low financial capacity to pay for response policies; variations in level of vulnerability; and divergent policy priorities such as prioritizing development over environmental protection. The familiar fault lines also hindered agreement: the North-South divide, energy producers versus energy consumers, countries at high and low risk, and ideological orientations toward the environment, in particular regarding the appropriate response to scientific uncertainty. Paterson and Grubb saw US policy as a major hindrance, concluding: "Finally, as a very large and technologically advanced country, the US has increasingly argued that it can adapt to climate change with relatively little cost."[31] In this assessment of the political climate heading into the Rio conference, here is one of only two mentions the authors make of adaptation to climate change, reflecting how low it had gone on the global climate policy agenda.

For his part, toward the end of his single term, President George H. W. Bush moved closer to his more pro-environment 1988 campaign position, although still not offering policies that would satisfy advocates of vigorous climate policy. After indicating he would not attend the summit, Bush did attend and spoke at the Rio conference, signing the

UNFCCC there. Bush's statement on the signing of the UNFCCC accepted the scientific consensus on the extent and causes of global warming, and, unlike nearly all his previous statements, it made no mention of the need to sustain strong economic growth. "But then," Scott Waldman and Benjamin Hulac write, "his administration did little to follow through with those pledges or to fight for something more concrete than a framework. In the end, Bush balked at the most meaningful part of the Rio gathering—setting specific emissions reduction targets."[32] The administration faced criticism for watering down the agreement, to which Bush responded that the United States would not accept commitments it could not keep. The ambivalence of the administration reflected the contending forces in American society. Significantly, the administration did not object to the climate science showing that warming was likely; instead, it emphasized the remaining uncertainties in the science, advocated a broad-front response that included adaptation, and called for no-regrets policies to take measures that would make sense regardless of remaining scientific uncertainty.[33] Notably, these limited responses to the climate challenge were also well represented in the IPCC report.

In preparation for the Rio Earth Summit, the IPCC produced its first Assessment Report following the format in use today: an analysis of the state of climate science, of the probable impacts on the natural environment and human interests, and of response strategies. Response strategies are grouped under mitigation (emissions reduction) and adaptation.[34] However, the IPCC's work, subject to the influence of governments, came under fire, perhaps marking the turn by 1992 toward much greater emphasis on mitigation than adaptation. Greenpeace criticized a UN conference that was discussing the IPCC's first major report for failing to specify sufficient measures to reduce GHG emissions.[35] Similarly, the Natural Resources Defense Council and various European diplomats noted the inconsistency between the warnings of the working group on the science of climate change and the lack of remedial measures in the report of the working group on response measures. The mixed messages of the IPCC report entered the policy dialogue and resulting agreements.

In addition to the IPCC's work, the General Assembly authorized the UNCED Preparatory Committee (PrepCom IV) to ready several documents for the upcoming summit: Agenda 21, the Earth Charter, and the Statement of Forest Principles. As an element of Agenda 21, PrepCom IV addressed the atmosphere, including climate change, as well as transboundary air pollution and ozone depletion. However, PrepCom IV mainly deferred to parallel talks by the Intergovernmental Negotiating Committee for a Framework Convention on Climate Change. Thus, while the draft of Agenda 21 included a chapter on the atmosphere, it was entirely bracketed and left for completion at UNCED.

The atmosphere chapter discussion was not taken up until the fourth week of PrepCom IV, when it focused on matters pertaining to lifestyle, consumption, transportation, and energy. The United States objected to the use of the term *lifestyle* because, the delegation said, it would be contrary to the Universal Declaration of Human Rights, which did not allow governments to determine citizens' lifestyles. Or, as President Bush said, the American way of life is not up for negotiation. Regarding the central concern of this study, PrepCom IV rarely mentioned adaptation to climate change. That lack of attention was reflected in the main product of the committee's work, Agenda 21. Chapter 9 on the atmosphere mentions the need to improve data and analysis, and to hold temperature rise low enough that ecosystems can adapt naturally. There was no notice given to the need to prepare society for higher temperatures with proactive measures. The chapter also cited the need to devote resources to capacity building, but this was not understood at the time to be about adaptation. Rather, capacity building was to be geared toward reinforcing developing countries' "technical, managerial, planning and administrative capacities to promote sustainable development and the protection of the atmosphere."[36] Capacity building as enhancing countries' resilience and adaptive capacity was not mentioned. The report on a preliminary meeting of the Intergovernmental Negotiating Committee for a Framework Convention on Climate Change noted the need for assessments of adaptation efforts, but the point was not elaborated.[37]

Consequent to these extensive preparations, at the 1992 Rio conference, governments agreed to the UN Framework Convention on Climate Change, by which time mitigation had entirely eclipsed adaptation. The stated purpose, presented in Article 2 of the UNFCCC,[38] is to stabilize the concentration of greenhouse gases in the atmosphere below the level that would cause "dangerous anthropogenic interference with the climate system." That level, as noted in Chapter 1, was thought to be 2°C above the preindustrial level. The statement of purpose mentioned adaptation, specifically that the international community should take action quickly enough "to allow ecosystems to adapt naturally to climate change," and subsequent articles in the UNFCCC address the adaptation of human systems. But the statement of purpose is squarely focused on mitigation. Mitigation was expected to give natural systems the time they needed to adjust to higher global temperatures. When social adaptation is noted, it is clear that negotiators expected the costs would fall largely on developing countries, and perhaps assistance would have to be provided. Developed countries, it was expected, would have sufficient resources to manage their own adaptation needs.

Presumably, more effective mitigation would reduce the need for adaptation, which would help explain the elevation of mitigation to the top policy concern. However, the weakness of commitments to mitigate climate change was already evident in the UNFCCC. Negotiators at the 1992 Rio Earth Summit had settled on three areas for international agreement—forests, biodiversity, and climate change—and all three major initiatives had been weakened, primarily to meet US objections to the proposed agreements. Writing on the eve of the Earth Summit, Katrina Brown observed, "After much negotiation, a convention on climate change has been provisionally agreed, but only after concessions were made to the world's biggest polluter, the United States, which its critics say has watered down the convention so much that it is almost meaningless."[39] Thus, the flaws of mitigation policy were written into the first major international agreement to address the problem, setting the stage for the later rise of adaptation on the agenda.[40]

The upshot. The common understanding of the development of climate policy assumes that adaptation was always a minor theme, only becoming a major element of the agenda when it had become obvious that mitigation would fall short of averting significant impacts on human society. Indeed, the Adaptation Committee observed:

> When the Convention entered into force in 1994, the primary focus was on greenhouse gas mitigation, and the idea of adapting to anthropogenic climate change struggled to gain legitimacy alongside mitigation. In some cases, the discourse surrounding adaptation implied that it betrays an arrogant faith in the capacity of natural and human systems to adapt, and that it could siphon attention and energy away from the urgent problem of reducing emissions.[41]

Yet as we have seen, the scientific meetings and reports emerging before political leaders took center stage on the climate issue anticipated considerable that adaptation would be necessary, feasible, and probable. Indeed, early discussions tended to assume that human societies would adjust in various ways to an inevitably warming planet, just as they had adapted to climate variability many times in the past. Further research on climate was justified in large part because better predictions would enable better adaptations in anticipation of probable impacts to come.

This all changed by 1992 and the negotiation of the UNFCCC, when mitigation was forefronted and adaptation, while mentioned, was not regarded as a central topic. By late 1990, to counter resistance to limits on emissions, the first political statement by a large group of scientists strengthened their advocacy of mitigation, and attention to adaptation fell commensurately.[42] A concise history of climate policy posted to the UN's website does not even mention adaptation being part of the UNFCCC, nor does it note that it was a significant element of prior negotiations leading to the agreement.[43]

Various academics have noted that this denigration of adaptation was due, in part, to the sense that opponents of strong mitigation policies were arguing large reductions of GHG emissions were not needed if society could adapt to new climate conditions. To the extent that such an argument found resonance among policymakers, especially in Washington, it would make political sense for advocates of aggressive climate policy to act as if mitigation were the only viable option. And so, they did. However, this overlooked the transformative potential of adaptation. For measures short of transformation to suffice, planetary warming cannot impose conditions exceeding the adaptive capacity of civilization as we know it, a proposition that is now questionable. As doubts about mitigation policy rose, attention to adaptation increased.

Marrakesh 2001 COP7

In less than a decade, one that saw the negotiation of the important Kyoto Protocol to the UNFCCC, the inadequacy of mitigation measures had become evident, and policymakers began to raise the salience of adaptation.[44] The changing attitudes toward adaptation can be traced through two IPCC reports and the negotiations at the 2001 COP7 held in Marrakesh.

Second Assessment Report 1995. Why focus on IPCC Assessment Reports? It is important to bear in mind that the IPCC's ARs are not just more writings in a deluge of studies and policy documents. The ARs are much-anticipated and widely read summations of the state of knowledge. The IPCC reviews a vast body of literature in scholarly sources, which it then summarizes, analyzes, and presents with an eye to policy relevance.

This work influences scholarship and policy as nothing else does. Certainly, the full scope of scholarship is broad and encompasses dissenting views, but if a middle position exists that reflects a rough consensus among scientists and social analysts, the IPCC portrays it, albeit employing somewhat cautious language to do so.[45]

In 1995, the IPCC released the Second Assessment Report (AR2) reporting the products of three working groups, Working Group I, Working Group II, and Working Group III. Working Group I was charged with assessing the scientific basis for the climate challenge, and it affirmed the high likelihood that human activities would have unprecedented effects on the global climate. Certainly, any human influence on Earth's climate, no matter how slight, would be unprecedented, but the IPCC meant that the impact could have serious consequences for human society and human interests.[46] Yet the policy recommendations emerging from AR2 remained modest. Working Group III assessed mitigation options, including emissions reductions and GHG removal.

Working Group II's product is the most pertinent here, as it was charged with synthesizing research on the impacts of climate change on the ecology and society, adaptation measures, and options for mitigation of climate change.[47] The report is organized mainly according to different sectors of the ecology, economy, and society. Part II, "Assessment of Impacts and Adaptation Options," covers forests, deserts, mountains, coastal areas, and other ecosystems, and it assesses impacts and options for human settlements, infrastructure, agriculture, human health, and others. The significance of global warming for a given sector, said the report, depends on how sensitive, adaptable, and vulnerable it is to climate change. Each chapter was to explain what global warming would do to the sector in question and to identify how to adapt to rising temperatures in that area. Yet in some chapters, identification and elaboration of adaptation options is left out entirely, or it is treated superficially. This reflects, in part, the lack of research on adaptation at that time, as the authors state. It would also follow from the sheer lack of adaptation options; this barrier is stated explicitly regarding extensive protection of wetlands, for example.

The general tenor of the report's observations on adaptation is technocratic and managerial. Forests would likely require human interventions to assist natural migration, create reserves, and reduce other stresses on forests such as pollution and harmful land use practices.[48] Rangelands could benefit from adaptive strategies to "include the preservation of an extensive spatial scale in management units and the development of viable marketing systems," which sounds a lot like emphasizing large-scale agribusiness. Broadly, "Intervention—in terms of active selection of plant species and controlled animal stocking rates—is the most promising management activity to lessen the negative impact of future climate scenarios on rangelands."[49] Such adaptation as could be attained would rely on better-suited animals and crops, diversification, and more efficient use of water.

For low-lying islands and coastal areas, high costs come into play, and there will inevitably be trade-offs of "environmental, economic, social, and cultural values."[50] The recommended answer to these potentially profound trade-offs was "a flexible coastal management strategy at all timescales, that incorporates and integrates both short-term and long-term goals."[51] Creating and implementing such a plan would require effective leadership in initial planning, appropriate institutional arrangements, enhanced technical capacity, and management instruments that can encourage stakeholders to comply with the goals and objectives of the plan. It was expected that changes to institutional arrangements would mean primarily strengthening ties among existing global, regional, and national policymaking bodies. Clearly, adaptation policy was expected to occur within predominant institutions and in line with existing practices and worldviews.

Regarding industry, energy, and transport, the report notes that the "literature on adaptation and adaptation policies is particularly sparse."[52] Nevertheless, the report suggests that adaptation will depend on the duration of human-built infrastructure, with short-lived assets turning over quickly enough to change as climate conditions change, while long-lived assets could be badly suited to future conditions. The answer to long-term challenges is "the development of climate guidelines" and improved monitoring.[53] Human settlements would have to adapt to rising temperatures and consequent climate changes. The most vulnerable were expected to become more vulnerable as temperatures rise. However, adaptation would occur as people move, change occupations, install air conditioning, and the like as conditions change. Policy-directed responses could include appropriate land tenure rights, immigration laws, and a variety of policies that would improve conditions in urban areas apart from climate change. Transformation of existing development pathways was not on the agenda.

For human well-being, no sector matters more than agriculture, and here the prognosis was optimistic, finding "far greater potential for global agriculture to adapt to changing climate than earlier studies."[54] The responses that showed promise were "technological," including adjusting to changing growing seasons, planting different crop varieties or species, developing new varieties, improving water management, and using appropriate fertilizers. Institutional barriers to the adoption of such technological solutions were identified, emphasizing in response the need for planning, management, policies, and programs. Water management was also understood in technocratic and managerial terms; adaptation would entail new investments, better operation of existing systems, proper maintenance, and modifications of regulatory frameworks. The report concludes, "We will be able to adapt—though at some cost economically, socially, and ecologically."[55] Regarding forests, recommendations recognized the long time frame for forest management, meaning that a prudent policy orientation would prepare for severe undesirable impacts with a flexible set of options.[56]

For the financial sector, the report recommended increased recognition by financial managers that climate change could affect future prospects, and that this "could require institutional change,"[57] meaning modifications of regulations, industry practices, and incentives. To be effective, industry, governments, and international bodies would need to cooperate on standards, regulations, and governing structures. No notion of fundamental transformation of the global financial system was mentioned. Human health is vulnerable to many impacts due to climate change, and the "adaptive options," stated the report, include greater availability of medical care; management and planning for environmental change and disasters; technology that protects human health such as water purification; public health education; and education and training for health professionals.[58] Such recommendations would be useful regardless of climate change.

Several sectors saw either no explicit discussion of adaptation or comments that indicated adaptation would be of limited value. Regarding human-induced desertification, the only remark was that it could overwhelm efforts to contain the effects of climate change on deserts. No adaptation measures were identified for mountain regions, other than monitoring and research; the same can be said of the cryosphere, and hydrology and freshwater ecology. Responses to protect wetlands might be feasible at the local level, but "for regional and global functions (e.g., trace-gas fluxes and carbon storage), there are no human responses that can be applied at the scale necessary."[59] Adaptation to climate change for the oceans would likely be overwhelmed by the long-standing mismanagement of ocean resources, which would have to be corrected before effective adaptation would be possible.

Regarding institutional factors in general, the report stated, "The efficacy and cost-effective use of adaptation strategies will depend upon the availability of financial resources, technology transfer, and cultural, educational, managerial, institutional, legal, and regulatory practices, both domestic and international scope."[60] Moreover, in line with the view still expressed today, it asserted that developing countries are likely to be more vulnerable than developed countries. Nonetheless, the degree of institutional change implied by the report's recommendations was limited. Changes to institutional arrangements were to be understood as, for example, creating and strengthening agencies and funds that support adaptation. There was no indication in AR2 that adaptation entails transformation of broad social and economic institutions, including the state system and neoliberal global capitalism. The approach was entirely managerial and technocratic, assuming that extant global and national social, political, and economic institutions would carry on more or less as always. A recent study notes, "A proposed technocratic solution avoids political conflicts—it requires no change in behavior, rather only improved conditions for technical innovations, their commercialization and use."[61] In AR2, this orientation was predominant.

Third Assessment Report 2001. In the Third Assessment Report (AR3), Working Group II was again charged with "assessing available scientific and socioeconomic information on climate change and its impacts and on the options for mitigating climate change and adapting to it."[62] The preface to the thousand-page volume noted that it differed from the previous report in undertaking a regional assessment of climate impacts and policy options, and it included a separate chapter on adaptation and sustainable development. In addition, AR3 sought to address issues that cut across sectors and in other working group reports. Mitigation options had been discussed in Working Group II's contribution to AR3, but that topic was largely left to Working Group III in AR3. "The value of adaptation measures to diminish the risk of damage from future climate change, and from present climate variability, was recognized in previous assessments and is confirmed and expanded upon in the new assessment," stated the preface.[63] Unlike AR2, which focused on technocratic responses to climate impacts, AR3 suggested that equity is important to successful adaptation.

AR3 stated that many natural systems (e.g., retreat of glaciers, earlier emergence of insects) were already showing changes due to rising temperatures, at that time about 0.6°C above preindustrial levels. Natural systems, said AR3, can have limited adaptive capacity and thus "may undergo significant and irreversible damage."[64] The report identified water resources, agriculture, forestry, coastal and marine zones, human settlements, energy, industry, financial services, and human health as sensitive human systems. As before and since, developing countries were said to be more vulnerable and less able to adapt than developed countries. AR3 also suggested the possibility of tipping points that could lead to large-scale, irreversible changes such as slowing of Atlantic Ocean circulation, large reductions in ice sheets, and positive feedbacks that would lead to much greater, faster temperature increases than would occur only from human sources of GHG. Although considered low probability, such amplifications of climate change could outstrip the adaptive capacity of both natural and human systems.

When AR3 was published, important states had negotiated the Kyoto Protocol to the UNFCCC. The United States was notably absent from the list of ratifying countries. President Bill Clinton had signed the agreement, but President George W. Bush opted not to honor that signature or to accept the limits on emissions required by the agreement. The US Congress agreed with the president on this, and the lack of US participation, as well as obvious signs that mitigation by other countries was falling short, indicated that temperatures would rise. Consequently, AR3 asserted, "Adaptation is a necessary strategy at all scales to complement climate change mitigation efforts."[65]

The report outlined an expanded version of no regrets in that policies to adapt to climate change would also alleviate pressures from population growth, resource depletion, and poverty affecting primarily the poorer parts of the world. Presumably, alleviating poverty would increase a society's adaptive capacity, reducing vulnerability to climate change while also serving other goals of sustainable development. This includes the ethical imperative of achieving greater equity, in part justified by recognition that poor countries and regions would suffer the greatest impacts while having the least capacity to respond. On this account, equity, development, sustainability, and adaptability go hand in hand. The key was to include consideration of climate risks in national and international development planning and programs, later termed *mainstreaming*. Thus, beyond identifying technical and managerial responses to specific risks, AR3 suggested integrating climate adaptation with the broader development agenda that had prevailed in world affairs since at least the 1960s. Further research on the requisites for adaptation was required, noted AR3, for it was "expected to be useful for identifying successful strategies for enhancing adaptation capacity in ways that can be complementary to climate change mitigation, sustainable development, and equity goals."[66]

Drawing on the work of Working Group I, AR3 outlined serious consequences for human society of rising global temperatures. The regional focus, which had not been possible previously due to lack of adequate research, showed that some areas would likely suffer the greatest impacts, notably in sub-Saharan Africa, Latin America, and Asia. This, said AR3, called for greater attention to the adaptation needs of vulnerable communities, regions, and countries. However, said the report, attention "has focused largely on mitigation issues at the global or regional scale and only secondarily on issues of impacts, vulnerability, and adaptation."[67]

Despite gaps in available research, the authors felt sufficient confidence in the state of knowledge to assert that adaptive capacity depends on "access to resources, information and technology, the skill and knowledge to use them, and the stability and effectiveness of cultural, economic, social, and governance institutions."[68] At first glance, this implies that responding to the worsening effects of rising temperatures might require social, political, economic, and cultural measures commensurate with the challenge, which later reports would term *transformational change*. Yet AR3 did not suggest sweeping institutional transformation, although the idea might have been implicit in the logic of adaptation to climate change even then. To the contrary, AR3 adopted the opposite tack, asserting that effective adaptation can only occur when it is in line "with existing policy criteria, development objectives, and management structures."[69] Implementation of adaptation measures is only likely, said AR3, when it works with other policy frameworks concerned with nonclimatic challenges. That is, adaptation must accommodate itself to such policy criteria as striving for economic growth, development objectives such as reinforcing neoliberal models, and management structures such as corporate boards and government agencies—tempered, of course, by the sustainable development agenda. Indeed, "*Activities required for enhancement of adaptive capacity are essentially equivalent to those promoting sustainable development.*"[70] Although sustainable development was a central concern of the 1992 Rio Summit, AR2 did not highlight that broad objective; in AR3, it is all but synonymous with adaptation.

The problem is that this lands the adaptation problem in the middle of an intense debate over the meaning and implications of sustainable development. Sustainable development can imply anything from relying on the profit motive to instituting robust global regimes and domestic legislation to defend environmental values.[71] Beyond this, critics note that the concept of sustainable development privileges economic growth, viewing environmental sustainability as a necessary condition for further development, not a value in its own right. "That is, built into the notion of sustainable development," Arran Gare asserts, "is a bifurcation where what really matters is the development of the economy, and the ecosystems of which humans are part are considered in terms of whether they are sustainable under the stress of the developing economy."[72]

Clearly, in declaring that adaptation must work with existing priorities and organizations, AR3 viewed sustainable development as carrying on business as usual while implementing appropriate regulations and incentives to cope with environmental problems. The report asserted that climate adaptation and equity goals could be pursued at the same time through "initiatives that promote the welfare of the poorest members of society—for example, by improving food security, facilitating access to safe water and health care, and providing shelter and access to other resources."[73] This folds adaptation neatly into the broader development agenda, which implicitly assumes that providing basic needs to the least advantaged amounts to an act of benevolence and charity by those better off, helping the poor to correct countless policy mistakes and institutional failures that put them in their sorry condition. But whether this orientation was adequate to the climate challenge or justifiable in terms of social aspirations for justice, equity, and democratic rule was and remains subject to debate. Indeed, later IPCC studies increasingly move away from this implicit endorsement of the institutional status quo, although the emphasis on sustainable development is even stronger in the most recent reports than in AR3.

In sum, AR3 differed from AR2 in important ways. While AR2 was technocratic in approach, identifying specific remedial measures to reduce vulnerability and increase resilience, AR3 addressed institutional questions more directly, in particular suggesting how adaptation could work with sustainable development. However, AR3 insisted that adaptation policies and programs would have to fit with the broader institutional framework of world affairs. Without calling it by name with an accurate, explicit term,[74] the thrust of the document was that adaptation in particular and sustainable development more broadly would have to conform to the requisites of global capitalism. Adaptation would not be transformative; instead, it would be a component of the reformist approach of sustainable development.

Fourth Assessment Report. Prior to COP7, the IPCC released the Fourth Assessment Report (AR4), which included the findings of Working Group II on responses to climate change. Echoing AR3, the IPCC asserted that adaptation to climate change should work in conjunction with sustainable development, a theme that would become stronger in subsequent IPCC assessments. Reiterating the findings of previous reports, it identified influences on adaptive capacity: "A society's productive base, including natural and man-made capital assets, social networks and entitlements, human capital and institutions, governance, national income, health and technology," as well as development policy. Barriers to the proper implementation and effectiveness of adaptation measures encompass financial, technological, cognitive, behavioral, political, social, institutional and cultural dimensions.[75]

The IPCC advised a risk management approach to coping with the effects of a warming world. AR4 envisions an "iterative risk management process" for both mitigation and adaptation, "taking into account actual and avoided climate change damages, co-benefits, sustainability, equity, and attitudes to risk." These considerations would apply to five "areas of concern" outlined in the previous report to guide policy makers: risks to unique and threatened systems; risks of extreme weather events; distribution of impacts and vulnerabilities; aggregate impacts; and the risk of large-scale singularities (tipping points). Adaptation, the IPCC declared, would be of little avail in a world of unchecked GHG emissions, in which climate change "would, in the long term, be likely to exceed the capacity of natural, managed, and human systems to adapt."[76] An effective response would require both mitigation and adaptation.

The question is whether the suggested approach would be sufficient to cope with the problem. In AR4, the general tenor is that adaptation is about technical management to achieve risk reduction, and it would take the form of planning by experts who consult with local and Indigenous peoples. The authors noted the positive cost-benefit ratios of timely, appropriate adaptation. Like other studies, it also recommended that adaptation be mainstreamed into development planning.[77] However, AR4 makes little mention of the ethical dimensions of adaptation such as polluter pays and common but differentiated responsibilities. Equity is considered primarily regarding possible influences on the efficacy of adaptation, rather than as a value to be pursued in its own right. AR4 posits a range of adaptation options, all well within the framework of existing institutions: "The array of potential adaptive responses available to human societies is very large, ranging from purely technological (e.g., sea defences), through behavioural (e.g., altered food and recreational choices, to managerial (e.g., altered farm practices) and to policy (e.g., planning regulations)."[78]

While the report assesses this range to be large, it leaves out deeper transformations of economic and social practices, such as ending corporate control over the productive capacity of society, which some would say has caused the ecological crisis. The IPCC acknowledges in many places the severe barriers in nature and society hindering adaptation, but it does not analyze or confront these barriers, other than to say they are not well understood, and developing countries need greater resources and adaptive capacity building.

Evidently, the IPCC preferred to avoid discussing some well-understood aspects of the economic barriers to adaptation policy; namely, corporate influence on policy, class relations, and the broad attributes of a growth- and profit-oriented system that would tend to define adaptation as whatever furthers the continued accumulation and concentration of capital in ever fewer hands. Rather than contemplate the kind of systemic transformation that might be required to enact effective, equitable adaptation to climate change, the IPCC found hope in the development of National Adaptation Programmes of Action (NAPAs) by the least developed countries (LDCs), which, the IPCC suggested, are "attempting to overcome" the immense institutional barriers even to technical, managerial, incremental adaptation, let alone the transformational change later suggested in AR5. The barriers and limits to adaptation noted in AR4's Chapter 17 do not include global capitalism, neoliberalism, state sovereignty, the threat of military conflict, or any of a number of other impediments to successful adaptation to climate change.[79] Brief mention of the less helpful aspects of the state system is made in the subsequent chapter on the interrelationship of adaptation and mitigation: "For example, particular development path scenarios that give rise to very high emissions might also be associated with a slower growth, or even a decline in the determinants of response capacity. Such might be the case in scenarios with high degrees of military activity or a collapse of international co-operation. In such cases, climate change impacts could increase, even as response capacity declines."[80]

Nevertheless, having nodded toward the broad institutional framework that could thwart even the best-intended climate policy, the authors suggest linking sustainable development to climate policy through "productive partnerships with the private, public, non-governmental and research sectors"; that is, with little challenge to the global institutions that have brought the world to this situation. Indeed, the authors endorsed the trend toward folding climate policy into the development agenda, which meant linking it to development assistance and thus the various problems attending that manner of addressing global challenges.

The 2001 Marrakesh Conference of the Parties

The International Institute for Sustainable Development (IISD) report on the 2001 Marrakesh Conference of the Parties (COP7) took note of a few agenda items on adaptation, but the focus of the conference was clearly on mitigation. The main topic was to move toward ratification of the Kyoto Protocol, which required reaching agreement on how to assess compliance and commitments to GHG reductions. Negotiators had failed to agree on critical implementation questions in the two-part COP6, and the United States announced in advance of COP7 that it found the Kyoto Protocol unacceptable. Consequently, COP7 was consumed with efforts to reconcile contenders and establish rules for GHG reductions, with some hope that the United States could be brought into the fold.[81]

To be sure, parties did draw attention to the vulnerability of developing countries and the consequent need for adaptation funding. Draft decisions called for support for capacity building, and to provide guidance to the Global Environmental Fund (GEF) for establishing pilot and demonstration projects that would show how to go from planning and assessment to implementation of adaptation projects. But the underdeveloped state of adaptation policy implementation led to emphasis on research, information sharing, and monitoring. The need to develop methodologies for assessing adaptation was noted, without which progress on adaptation would be hindered.[82]

Despite such limitations, Marrakesh COP7 is considered to be a key moment for adaptation policy, when adaptation rose in importance. The Global Environment Fund reiterated its support for adaptation funding, the IPCC identified barriers to adaptation that would need redress, and many countries "underscored the need for capacity building, adaptation and technology transfer."[83] Delegates had agreed on a "special climate fund and/or adaptation fund" and called for other sources of bilateral and multilateral funding to further adaptation and related activities. The new fund would be expected to provide support for the preparation of NAPAs. In closing remarks, many countries reemphasized the need for adaptation, along with capacity building and technology transfer. The most vulnerable countries, including Bangladesh and other low-lying states, and the small island states, urged rapid dissemination of funds for adaptation. The delegate from Thailand called for "a balance between adaptation and mitigation activities,"[84] marking the salience of adaptation at a conference that was otherwise absorbed with making the Kyoto Protocol on mitigation effective.

COP7 occurred after several contentious meetings on the meaning and implementation of GHG reductions mandated in the Kyoto Protocol. Some of the developed and fossil fuel–producing countries showed no interest in a strong framework for emissions reductions, and they used their leverage to extract significant concessions. The European Union (EU) and China/Group of 77 (G-77) sought to strike a deal that would ensure enough ratifications for the Kyoto Protocol to enter into force, which required conceding to the demands of the bloc of countries wanting to dilute the agreement's requirements, called the Umbrella Group. Given the Umbrella Group's orientation toward reliance on and profit from fossil fuels, one might expect that its members would lay emphasis on adaptation as a sufficient remedy for temperature rises, for them to insist that adaptation and mitigation could be balanced in such a way that oil, gas, and coal would remain a major source of energy. From this perspective, placing more of the weight on adaptation would reduce the need for strong emissions reduction requirements.

However, the record does not indicate that the Umbrella Group or any individual country presented such an argument. James Ford observes that developed countries steered away from discussing adaptation, in part because "there was considerable faith in the ability of mitigation to be effective in tackling climate change."[85] In addition, some developed country governments avoided the topic because acknowledging the need for adaptation (including funding it) amounted to "accepting human responsibility for climate change," which would in turn lead to calls for liability and compensation to those harmed. Not until the scientific consensus on the effects of human activity on the climate had become overwhelming would developed countries engage in discussion of adaptation and its financial implications.

Instead, when adaptation came up, it was raised by vulnerable developing countries wanting funding. Malaysia, for instance, "emphasized the vulnerability of developing countries and those in poverty and, with the Republic of Korea, supported consideration of adaptation measures."[86] In the closing general meeting, delegates from various developing countries made a plea for more attention to adaptation, and for the funding to be provided for it. Prominent developed countries such as the United States, and large oil producers such as Saudi Arabia, did not, according to reports, contribute to this discussion, while their remarks on mitigation are noted. Portugal was the only EU member recorded to have spoken on the topic, observing that adaptation could promote sustainable development and intergenerational equity.

In UN documents detailing the outcomes of the conference, adaptation receives considerable attention. But the context is nearly always a call for special attention to the needs of highly vulnerable developing countries. Regarding capacity building, technology transfer, and financial arrangements, the documents note the importance of the activity for assisting adaptation in developing countries. To be sure, these measures are intended first to enable developing countries to mitigate GHG emissions. But they are also expected to enhance adaptation, although at that time the main recommendation was for more research and information efforts.[87] The general orientation is familiar: developing countries are the most vulnerable to climate change, lack the capacities and resources to adapt effectively, and will require financial assistance to enable them to do so. Even in 2001, participants were pointing to impacts that had already been felt and the likelihood of more profound effects in the future. In the final documents and in the available records of the debates and discussions, nobody asserts that adaptation would obviate the need for mitigation. Instead, adaptation seems to have been understood more as a regrettable necessity in the pursuit of sustainable development.

Outside official circles, the notion that adaptation could trade off with mitigation, and indeed might be the preferable option, did find adherents. To cite a prominent voice, Bjorn Lomborg observes that global warming likely would cost $5 trillion by the end of this century. However, he continues, "Despite the intuition that something drastic needs to be done about such a costly problem, economic analyses clearly show that it will be far more expensive to cut carbon-dioxide emissions radically than to pay the costs of adaptation to increased temperatures."[88]

Lomborg suggests that global warming was being used as a "springboard for other wider policy goals"; namely, remaking global civilization, whereas "global warming is not anywhere [near] the most important problem in the world." Rather, "what matters is making the developing countries rich and allowing the citizens of developed countries even greater opportunities."[89] Economic growth, then, is the appropriate response to climate change, not even requiring a rapid reduction in burning of fossil fuels. No doubt, many interested parties found this argument congenial. However, the impact of this point of view on COP7 would have been considerably diminished because President George W. Bush decided that the United States would not take an active part in the negotiations.

The creation of the Adaptation Fund is considered a notable achievement of COP7. The Adaptation Committee's (AC) history of adaptation policy posits an evolution of thinking from information gathering to "planning and pilot implementation" initiated at COP7,[90] and the Adaptation Fund (AF) was an important part of that transition. The AC asserts that, with the release of the IPCC's third report in 2001, it had become clear that adaptation would be required; the question had moved from whether adaptation was necessary to how to do it. The AC noted that limited evidence indicates adaptation funding, when provided, makes a positive difference, but for that to happen requires a series of prior steps to ensure that adequate funding reaches vulnerable communities.[91] Complexities in implementation have hindered adaptation finance,[92] which still falls short of pledges to aid impacted countries and communities.[93]

The AF was established under Article 12.8 of the Kyoto Protocol to the UNFCCC. The article states that the parties to the protocol will provide financial assistance to countries that are especially vulnerable to climate change and most in need of funding for adaptation measures."[94] For some years after the AF became operational, the focus was on providing funding and technical assistance to developing countries to produce National Adaptation Plans (NAPs).

Establishing the Adaptation Fund might have marked an increasing emphasis on adaptation, but it took some time for the fund to become operational, which did not happen until after the Kyoto Protocol entered into force. Also, important parties to the UNFCCC had not accepted the Kyoto Protocol, including the United States. In 2005, parties to the protocol adopted guidelines for the AF. In 2007, governance arrangements were agreed on, including interim reliance on the World Bank and the Global Environmental Facility for support services. Not until September 2010 did the AF approve its first two projects, totaling $14 million.[95] Thus, while COP7 signaled rising concern for adaptation, the lack of urgency for operationalizing the AF indicated that mitigation remained the top priority. However, other initiatives were under way, somewhat strengthening the turn toward adaptation. The work program adopted in 2001 included support for development of National Adaptation Programmes of Action, assisting LDCs in identifying and communicating their most urgent adaptation needs. In addition, the LDC Experts Group was established to advise countries on how to produce a NAPA, and funding for the work was provided through the Least Developed Countries Fund (LDCF). The COP created a mechanism for information sharing on adaptation experiences, supported by the Special Climate Change Fund and the AF.

The difficulty of distinguishing adaptation funding from general development funding stood as one of the barriers to the implementation of adaptation finance. Developing countries wanted additional funding for measures and projects for adaptation, but mainstreaming adaptation would make it difficult to ascertain whether more funding had been provided. "Such additionality is a key requirement for recipient countries in any international adaptation regime," wrote Hans-Martin Füssel, "because their vulnerability to climate change would not be improved by a mere relabelling of development assistance."[96]

Buenos Aires 2004: COP10

By 2004, parties attending COP10 in Buenos Aires placed emphasis on adaptation, in light of evidence that climate change was already having adverse impacts. In meetings of the Subsidiary Body for Scientific and Technological Advice (SBSTA), some parties called for more adaptation research, encouraging public participation, more funding, adaptation technology transfer, attention to Indigenous knowledge and practices, and stronger ties to other international agreements. A work program, mainly focused on weather disasters, was proposed that would encompass integration of adaptation into sustainable development. However, the United States pushed back by asserting there was no proof that extreme weather events were connected to climate change, and some parties insisted that the costs of responding to climate change be balanced against the benefits of adapting. In sum, "reaching an agreement on the "adaptation package" was therefore an arduous struggle and the urgency to respond to the impacts of climate change was not evident during the negotiations."[97] The parties agreed on a statement of support for implementing adaptation measures while also pursuing further research on the costs of responding to climate change.

The least developed countries had sought for two years to obtain a decision for "full cost funding of adaptation through the Global Environment Facility" and failed. In addition to continuing skepticism from such countries as the United States about the link between rising CO_2 concentrations and observed weather events and trends, it was also impossible to disentangle adaptation from the broader development process. The report on COP10 noted, "Adaptation is an integral part of development, and as such, no project directed at adaptation will fall squarely within the scope of the UNFCCC, but will rather have components that include other aspects of development such as disaster preparedness, water management, desertification prevention, or biodiversity protection."[98]

In practice, this meant that funding decisions were complex to the point of intractability. The rules stated that a given project could receive funds targeted at adaptation covering only that share of the total cost, while other sources of funding would be required for the economic development aspect of the project. Consequently, approvals required difficult calculations as well as facing hurdles in finding cofinancing, rendering the LDCF all but inaccessible to countries eligible for and seeking adaptation funding. LDCs did not have the bargaining power to overcome the lack of urgency that large fossil fuel users and producers expressed.

The following year, COP11 addressed adaptation in similar terms. After years of secondary status, "adaptation has become a key piece of the response to climate change—so much so that both COP 8 and COP 10 were dubbed 'the Adaptation COP'—particularly as the effects of climate change become more evident."[99] But this growing awareness produced incremental improvements, at most, to adaptation policy. Parties endorsed prior decisions to establish work programs on adaptation, while developing countries called for immediate action on adaptation measures. However, disputes over funding continued, and more meetings and research were all that the conference could achieve. Parties noted that knowledge and experience with adaptation were expanding and stated that the response to climate change should proceed in conjunction with the broader development agenda. Of course, this formulation retained the difficulties for funding decisions due to the entanglement of adaptation with development.

Montreal and Nairobi 2005–2006: COP11 and COP12

According to the Adaptation Committee's unofficial history, COP11 in Montreal and COP12 in Nairobi marked significant moments in adaptation policy. Presumably, these conferences signaled the international community's turn toward sharing knowledge, filling gaps in existing understanding, and setting the stage for implementation of adaptation. The most prominent outcome in this regard was the Nairobi Work Programme (NWP), adopted in Montreal and further developed in Nairobi the following year. The AC states, "The objective of the Nairobi work programme has been to assist all Parties, in particular developing countries, to improve their understanding and assessment of impacts, vulnerability and adaptation, and to make informed decisions on practical adaptation actions and measures."[100] The NWP built on and furthered development of NAPAs, which the LDCs were expected to produce so that funding and action could be properly targeted.

Beyond the usual calls for more research and reporting, the NWP was to be a kind of clearinghouse of information on adaptation experiences, which could provide useful knowledge to others facing similar adaptation challenges. "It serves," says the AC, "as a technical bridge between Parties, constituted bodies, and non-Party stakeholders to share and disseminate knowledge and experience on all aspects of vulnerability and adaptation, and to generate partnerships to collaborate on adaptation."[101] Thus, while it might have facilitated projects and other activities, the NWP was not itself about implementing adaptation, whether that means stronger infrastructure, moving communities, shifting economic activity, or transforming institutions.

The IISD's contemporaneous report on COP12 did not mention the NWP by name, which might indicate that the main business of the conference was elsewhere. Yet the IISD noted the SBSTA had adopted a work program on adaptation that included enhancing research quality. Actions and deliverables encompassed such things as workshops, reports, and technical papers. The call for more study of the problem apparently did not arise from the usual developed country ploy to put off action. Instead, the G-77 and China urged greater efforts on this front, perhaps because the difficulty of separating adaptation from development had thwarted disbursement of money from the Adaptation Fund and other sources.[102] The NWP could facilitate production of NAPAs, required before funding would be allocated. Whether called the NWP or simply a work program of the SBSTA, the increased attention on information sharing seems to have reflected a desire to overcome the procedural hurdles facing those countries seeking adaptation funding. To the extent the NWP helped overcome those hurdles, it could be counted as forward progress on adaptation.

That said, the general tenor of COP11 and COP12 remained well within the ambit of the broader development agenda, supporting the institutional status quo. The conferences emphasized long-term action on climate change, looking beyond the end of the Kyoto Protocol's first commitment period, scheduled to end in 2012. With a nod toward advancing development "in a sustainable way," the focus was on "realizing the full potential of market-based opportunities."[103] The call to rely on market-based opportunities might have reflected despair at the prospects for collective action to cope with climate change, including the possible "perverse effect" of providing funds for adaptation that might reduce incentives to mitigate.[104]

The emphasis remained on mitigation, despite recommendations from some in academia, as well as developing countries, to increase attention to adaptation. This was in part due to the manner in which the UNFCCC had defined the problem; namely, in terms of how to limit the long-term consequences of rising GHG concentrations, while largely ignoring near-term consequences and policy responses. To that point, Roger A. Pielke Jr. argues that mitigation could have only long-term benefits: even assuming the strongest possible mitigation policies, adopted and implemented in good faith, mitigation would not have any appreciable effect on the amount of global warming or its effects for decades. Consequently, "the only policies that can effectively be used to manage the immediate effects of climate variability and change will be adaptive."[105] Yet despite such calls from academia for greater attention to adaptation (including scholars cited in Pielke's article), the policy focus shifted only slightly. The main change was to call on the broader development process to provide adaptive capacity for vulnerable societies. The next phase of climate policy did show a growing concern for adaptation.

Bali 2007: COP13

In September 2007, Secretary-General Ban Ki-moon convened a High-Level Event on Climate Change, at which he asserted that a new deal on climate should include "adaptation, emission reductions, climate-friendly technologies and the necessary financial architecture." Two months later, the IPCC's Fourth Assessment Report (AR4) noted the need for greater adaptation efforts. Presumably, the "taboo on adaptation" had been lifted, with growing recognition that the probable best response to the climate challenge was sustainable development.[106] In part, this was due to a shift in corporate attitudes, from disputing whether human-induced climate change was occurring to identifying "how firms should adapt to climate change."[107] No doubt, corporate opposition also abated to the extent that the term *adaptation for sustainable development* was defined as furthering economic growth and enhancing the productivity and efficiency of economic sectors likely to be affected by climate change.[108]

Subsequently, parties adopted the Bali Action Plan, an agreement on long-term action on the entire agenda of climate action. It called for "enhanced action on adaptation," encompassing "vulnerability assessments, prioritization of actions, financial needs assessments, capacity-building and response strategies, integration of adaptation actions into sectoral and national planning, specific projects and programmes, means to incentivize the implementation of adaptation actions, and other ways to enable climate-resilient development and reduce vulnerability of all Parties."[109] In addition, the Bali Action Plan called for risk management, disaster reduction strategies, economic diversification, and greater cooperation toward support for adaptation. It also urged greater emphasis on technology transfer and financial support for adaptation.

"The main focus in Bali, however," stated the IISD, "was on long-term cooperation and the post-2012 period, when the Kyoto Protocol's first commitment period expires."[110] That is, Bali included important policies such as finalizing the Adaptation Fund, but its main goal was to extend and strengthen the mitigation commitments of the Kyoto Protocol. The COP decided to address a range of issues, including support for implementation of adaptation actions, with particular concern for the most pressing needs of the most vulnerable countries; namely LDCs, Small Island Developing States, and African nations. Nevertheless, as is typical of climate negotiations, lofty aspirations are rarely met with concrete action, and developed countries have yet to fulfill their funding pledges on adaptation. Meanwhile, one of the four major elements of discussions was how to maximize market-based opportunities to cope with climate change, and climate policy in general was expected to work in conformity with the traditional development agenda.

Delf Rothe observes that the climate skeptics had moved toward adaptation as they turned away from challenging climate science. The skeptics, he writes, had asserted that climate research was like religious thought, predicting an apocalypse with little foundation in fact. But in early 2007, following the publication of the IPCC's AR4 and the Nobel Prize given to former vice president Al Gore and the IPCC, the opponents of strong mitigation policy began to claim instead that it was too late to achieve the 2°C goal and therefore, they argued that the focus should be placed on efforts at adaptation and the development of new technologies.[111]

Looking beyond Bali, negotiators hoped to provide a road map for a comprehensive, abiding agreement to be finalized at Copenhagen in 2009. Several parties urged "early action" on adaptation to improve access to funding and to address observed effects of climate change.[112]

Adaptation might have been the only positive to emerge from the chaotic, failed Copenhagen conference. According to Earth Institute scientist Stephen Zebiak, adaptation was "the key" to progress on climate policy.[113] "In fact," said Zebiak, "an agreement on adaptation is the one thing that could come out of Copenhagen." However, instead, the next major development in adaptation policy occurred at Cancun in 2010.

Cancun 2010, COP16: The Adaptation Framework

In 2009, the UNFCCC process fell into chaos at Copenhagen's COP15.[114] The main product of the conference was the Copenhagen Accord, which was not adopted but only "noted" in the final decision documents. Various factors in US domestic politics as well as the usual cleavages at climate conferences thwarted progress toward binding, ambitious mitigation measures. Some advances on financing climate action presumably constituted the main accomplishments of the negotiations. But the secretary of the UNFCCC did not include adaptation in a summary of the main tasks facing the COP, and mitigation was by far the focus of discussions. Indeed, at one point, China called for dropping all other items from the negotiations until a satisfactory agreement on emissions was reached. In the end, the Copenhagen Accord did little more than reiterate the usual calls for adequate action on adaptation, including that developed countries should provide the funding.[115] Consequently, the next major step in adaptation policy awaited the following year's COP16 held in Cancun.

"Expectations for Cancun were modest," observed the IISD, but many hoped for a balanced package of outcomes that would include agreements on adaptation.[116] The Adaptation Committee deemed that Cancun represented "moving towards comprehensive adaptation," embodied in the Cancun Adaptation Framework (CAF) and the establishment of the Global Climate Fund (GCF).[117] Further, 2010 marked an important moment in adaptation policy in that negotiators emphasized that adaptation must receive the same priority as mitigation, a commitment supposedly reflected in the CAF. Consequently, the question under consideration was how to integrate adaptation into a broader global agenda of social, economic, and environmental policies and actions.

The CAF invited parties to act on a number of fronts, a significant share being further research and information gathering. Beyond that, the CAF called for actions, projects, and programs, and for stronger institutional capacities for adaptation. To build resilience, the CAF suggested economic diversification, sustainable management of natural resources, technology transfer, and enhanced disaster risk reduction.[118] In addition, parties established a process for formulating and implementing NAPAs, which would help to overcome the hurdles to accessing adaptation funding.

Developing countries displayed a more ambitious approach to adaptation at Cancun than at previous conferences. This was reflected in deliberations in the various committees charged with drafting new agreements. Developing countries insisted that climate funding give equal priority to mitigation and adaptation, and they wanted a lot more money for both. For example, in the Ad Hoc Working Group on Long-Term Cooperative Action (AWG-LCA), "A request was also made to insert text requiring Annex I parties to contribute 6% of the Gross National Product (GNP) to finance mitigation and adaptation in developing countries and 1% of their GNP to support forest-related activities as 'repayment of their climate debt.'"[119] If enforced today, that would require a country the size of the United States to contribute $1.9 trillion per year, whereas in practice the world has yet to fulfill its pledge to provide $100 billion per year in climate funding. The more moderate proposal developing countries supported called for developed countries to contribute 1.5 percent of GDP, which today would demand over $300 billion from a US-sized economy alone. More broadly, the idea that developed countries had incurred liability for damages by relying so heavily on fossil fuels was gaining ground. Needless to say, developing country demands for high levels of climate funding were not successful.

Unlike the Copenhagen outcome, where the COP could only take note of the main agreement reached, the parties adopted the Cancun Agreement covering all aspects of climate policy, including mitigation and adaptation. It called for enhanced action on adaptation, meaning support for "implementation of actions aimed at reducing vulnerability and building resilience of developing country parties, taking into account the needs of those that are particularly vulnerable."[120] In addition to the Cancun Adaptation Framework, the COP created the Adaptation Committee and began work on loss and damage. Measures taken to advance adaptation were considered to be among the positive outcomes of COP16, an important meeting for redeeming the COP process after the debacle in Copenhagen.

Assessment: Climate Policy and Development

Reviewing literature on adaptation available in 2011, Katrina Brown observes, "In each of these cases, conventional development interventions are shown to make poor people more vulnerable to climate change and less able to adapt," and further, "climate change is viewed by most development practitioners as something that affects development, and from which development needs to be protected, rather than a symptom of unsustainable development."[121] This remark accurately reflects the tenor of adaptation in climate policy from the 1992 Rio conference to the formulation of the Cancun Adaptation Framework in 2010. Development provided the framework within which adaptation policy was discussed, in part because the development process required addressing ecological effects to proceed. More importantly, development was seen as the best answer to the problem. The practical upshot was that developed countries should provide greater funding to developing countries to further development and to mainstream adaptation into the development process. This tendency is in line with the global reorientation to market-based solutions to all problems advanced since the 1980s.

The climate challenge rose on the global agenda just as neoliberalism had reoriented the world toward market approaches to regulating the international political economy. Accordingly, climate policy in the 1990s and since has developed against the backdrop of neoliberalism. As adaptation became a more prominent part of the climate policy discourse, neoliberal approaches to adaptation were devised. The most extreme manifestation of this tendency might be the increasing emphasis on ecosystem services and the related concept of ecosystem-based adaptation (EbA).[122] The notion of ecosystem services leads to the commodification of nature, while EbA provides an appealing story to further the project of commodification. Similarly, developed countries would insist that the pledge of $100 billion per year for climate funding be met in large part from private sector involvement rather than the budgets of developed country governments.

The Washington Consensus, writes Dorothy Grace Guerrero, "demolished" the principle of equity that had been advanced in both the 1972 Stockholm conference and the 1992 Rio conference. [123] The same could be said of other important principles of climate policy, such as common but differentiated responsibilities, which has recognized the different levels of contributions to the problem and capacities to respond.[124] The diversion of adaptation policy from transformative potential to support for the status quo was present at the beginning, but it has become more salient as the shortcomings of mitigation policy have manifested.[125]

Notes

1. Lisa Dale, *Climate Change Adaptation: An Earth Institute Sustainability Primer* (New York: Columbia University Press, 2022), p. 1.

2. Peter Jackson, "From Stockholm to Kyoto: A Brief History of Climate Change," *UN Chronicle,* XLIV, no. 2 June 2007.

3. John W. Zillman, "A History of Climate Activities," *Bulletin* 58, no. 3 (2009): 141–150.

4. WMO (World Meteorological Organization), *World Climate Conference: Extended Summaries of Papers Presented at the Conference* (Geneva: Secretariat of the World Meteorological Organization, 1979), p. 3. Robert E. Munn and Lester Machta's paper, "Human Activities that Affect Climate," in this document outlines the options for mitigation that are still under discussion today, including alternative energy, carbon capture and storage, increasing biological sinks, and reducing non-carbon GHG emissions.

5. WMO, *Proceedings of the World Climate Conference: A Conference of Experts on Climate and Mankind* (Geneva: Secretariat of the World Meteorological Organization, 1979).

6. Evgenii K. Fedorov, "Climatic Change and Human Strategy," in *Proceedings of the World Climate Conference: A Conference of Experts on Climate and Mankind* (Geneva: Secretariat of the World Meteorological Organization, 1979), pp. 15–26.

7. W. Lawrence Gates, "The Physical Basis of Climate," in *Proceedings of the World Climate Conference: A Conference of Experts on Climate and Mankind,* World Meteorological Association (Geneva: Secretariat of the World Meteorological Organization, 1979), p. 112.

8. Robert M. White, "Climate at the Millennium: Keynote Address," in *Proceedings of the World Climate Conference: A Conference of Experts on Climate and Mankind,* World Meteorological Association (Geneva: Secretariat of the World Meteorological Organization, 1979), p. 3.

9. WMO, *Report of the International Conference on the Assessment of the Role of Carbon Dioxide and of Other Greenhouse Gases in Climate Variations and Associated Impacts (October 9–15)* (Geneva: World Climate Programme, 1985), p. 42.

10. Irving Mintzer, "Communication on Energy: Is the Heat On?" *Energy Policy* 16, no. 1 (1988): p. 72.

11. Irving Mintzer, "Living in a Warmer World: Challenges for Policy Analysis," *Journal of Policy Analysis and Management* 7, no. 3 (1988), p. 456.

12. IPCC (Intergovernmental Panel on Climate Change), *Climate Change: The 1990 and 1992 IPCC Assessments* (Canada: World Meteorological Organization [WMO] and UN Environment Programme [UNEP], 1992).

13. WMO, *Second World Climate Conference—Conference Statement* (Geneva: WMO, 1990).

14. WMO, *Second World Climate Conference—Conference Statement* (Geneva: WMO, 1990), p. 3.

15. Suresh K. Sinha, "Impact of Climate Change on Agriculture: A Critical Assessment," in *Climate Change: Science, Impacts and Policy. Proceedings of the*

Second World Climate Conference, edited by Jill Jager and Howard L. Ferguson (Cambridge: Cambridge University Press, 1991), p. 107.

16. K. Madhava Sarma, "Adaptation Measures," in *Climate Change: Science, Impacts and Policy. Proceedings of the Second World Climate Conference,* edited by Jill Jager and Howard L. Ferguson (Cambridge: Cambridge University Press, 1991), p. 138.

17. Yu A. Izrael, "Climate Change Impact Studies: The IPCC Working Group II Report," in *Climate Change: Science, Impacts and Policy. Proceedings of the Second World Climate Conference,* edited by Jill Jager and Howard L. Ferguson (Cambridge: Cambridge University Press, 1991), pp. 83–86, emphasis in original.

18. Wolf H. Weihe and Raf Mertins, "Human Well-Being, Diseases and Climate." In *Climate Change: Science, Impacts and Policy. Proceedings of the Second World Climate Conference,* edited by Jill Jager and Howard L. Ferguson (Cambridge: Cambridge University Press, 1991), p. 348.

19. This subsection relies on the public papers of George H. W. Bush, https://bush41library.tamu.edu/archives/public-papers/.

20. Leslie Roberts, "Academy Panel Split on Greenhouse Adaptation: Its Conclusion that the United States Can Adapt Relatively Painlessly to Global Warming Draws Two Vigorous Dissents," *Science* 253, no. 5025 (1991), p. 1206.

21. *New York Times,* (AP), "Panel Says Nation Has Ability to Adapt to Global Warming," September 7, 1991, p. 12.

22. William D. Nordhaus, "A Perspective on Costs and Benefits," *EPA Journal* 16, no. 2 (1990): 44–45.

23. William D. Nordhaus, "To Slow or Not to Slow: The Economics of the Greenhouse Effect," *The Economic Journal* 101, no. 407 (1991), p. 933.

24. Eugene Linden, "Op-Ed: The Economics Nobel Went to a Guy Who Enabled Climate Change Denial and Delay," *Los Angeles Times,* October 25, 2018.

25. Mick Hamer, "Plot to Undermine Global Pollution Controls Revealed," *New Scientist,* January 2, 2002.

26. David Orton, "Reflections on Rio," *Canadian Dimension* 26, no. 7 (1992), p. 17.

27. Kenny Stancil, "Oil and Gas Industry's Expansion Plans Decried as Attack on 'Livable Planet'" Common Dreams, November 10, 2022, https://www.commondreams.org.

28. IISD (International Institute for Sustainable Development), "Overview of the UNCED Process," *Earth Negotiation Bulletin,* 2, June 1, 1992.

29. Matthew Paterson and Michael Grubb, "The International Politics of Climate Change," *International Affairs* 68, no. 2 (1992), p. 293.

30. George Monbiot, "Our Leaders Had a Final Chance to Halt Climate Breakdown. They Failed Each and Every One of Us," *The Guardian,* November 18, 2022.

31. Matthew Paterson and Michael Grubb, "The International Politics of Climate Change," *International Affairs* 68, no. 2 (1992), p. 302.

32. Scott Waldman and Benjamin Hulac, "This Is When the GOP Turned Away from Climate Policy," E&E News, December 5, 2018. At the same time, a National Oceanic and Atmospheric Administration (NOAA) official was arguing that social factors would matter more than raw scientific facts, and human society's creativity and perseverance could prevail over difficult natural circumstances, which amounted to making a case for adaptation; Charles N. Herrick, "Science and Climate Policy: A History Lesson," *Issues in Science and Technology* 8, no. 2 (1991): 56–57.

33. C. Boyden Gray and David B. Rivkin Jr., "A 'No Regrets' Environmental Policy," *Foreign Policy* 83 (1991): 47–65.

34. IPCC, *Climate Change: The IPCC Response Strategies* (Island Press, 1991).
35. "Greenpeace Faults UN Panel," *New York Times,* August 28, 1990.
36. UNCED (UN Conference on Environment and Development), *Agenda 21: Programme of Action for Sustainable Development, Rio Declaration on Environment and Development, Statement of Forest Principles: The Final Text of Agreements Negotiated by Governments at the United Nations Conference on Environment and Development,* June 3–14, 1992, Rio de Janeiro, Brazil, chap. 9.
37. UN, Intergovernmental Negotiating Committee for a Framework Convention on Climate Change "Draft Conclusions by the Co-Chairpersons of Working Group I," A/AC.237/WG.I/L.13, August 23, 1993.
38. UNFCCC (UN Framework Convention on Climate Control) Article 2:

> The ultimate objective of this Convention and any related legal instruments that the Conference of the Parties may adopt is to achieve, in accordance with the relevant provisions of the Convention, stabilisation of greenhouse gas concentrations in the atmosphere at a level that would prevent dangerous anthropogenic interference with the climate system. Such a level should be achieved within a time frame sufficient to allow ecosystems to adapt naturally to climate change, to ensure that food production is not threatened and to enable economic development to proceed in a sustainable manner.

39. Paul Brown, "Earth Summit: Long and Troubled Road to Rio—Paul Brown Traces How a Meeting in Sweden 20 Years Ago Gave Birth to the World's Biggest Conference," *The Guardian,* June 3, 1992.
40. President George H. W. Bush signed the UNFCCC at the Rio Earth Summit and defended the US record on climate change. He claimed, "It is estimated that our plan will reduce annual net greenhouse gas emissions by 125–200 million tons below projected levels by the year 2000." See "Bush Lays Down Rio Gauntlet: Call for New Global Meeting," *The Guardian,* June 13, 1992. Although evaluating Bush's assertion would be tricky, the fact is that US emissions rose steadily and dramatically from 1992 until the global recession of 2008. See Climate Action Tracker at https://climateactiontracker.org.
41. UN Adaptation Committee, *25 Years of Adaptation Under the UNFCCC* (Bonn: UN Climate Change Secretariat, 2019), p. 11.
42. Marlise Simons, "Scientists Urging Gas Emissions Cuts," *New York Times,* November 5, 1990.
43. Peter Jackson, "From Stockholm to Kyoto: A Brief History of Climate Change," *UN Chronicle,* XLIV, no. 2, 2007.
44. David J. Ciplet, Timmons Roberts, and Mizan R. Khan, "The Politics of International Climate Adaptation Funding: Justice and Divisions in the Greenhouse," *Global Environmental Politics* 13, no. 1 (2013): 49–68.
45. Keynyn Brysse, Naomi Oreske, Jessica O'Reilly, and Michael Oppenheimer, "Climate Change Prediction: Erring on the Side of Least Drama?" *Global Environmental Change* 23, no. 1 (2013): 327–337.
46. IPCC, *Climate Change 1995: The Science of Climate Change. Contribution of WGI to the Second Assessment Report of the Intergovernmental Panel on Climate Change* (Cambridge: Cambridge University Press, 1996), p. xi.
47. IPCC, *Climate Change 1995: Impacts, Adaptations and Mitigation of Climate Change: Scientific-Technical Analysis. Contribution of Working Group II to*

the Second Assessment Report of the Intergovernmental Panel on Climate Change (Cambridge: Cambridge University Press, 1995).

48. IPCC, *Climate Change 1995: Impacts, Adaptations and Mitigation of Climate Change: Scientific-Technical Analysis. Contribution of Working Group II to the Second Assessment Report of the Intergovernmental Panel on Climate Change* (Cambridge: Cambridge University Press, 1995), p. 113.

49. IPCC, *Climate Change 1995: Impacts, Adaptations and Mitigation of Climate Change: Scientific-Technical Analysis. Contribution of Working Group II to the Second Assessment Report of the Intergovernmental Panel on Climate Change* (Cambridge: Cambridge University Press, 1995), p. 147.

50. IPCC, *Climate Change 1995: Impacts, Adaptations and Mitigation of Climate Change: Scientific-Technical Analysis. Contribution of Working Group II to the Second Assessment Report of the Intergovernmental Panel on Climate Change* (Cambridge: Cambridge University Press, 1995), p. 292.

51. IPCC, *Climate Change 1995: Impacts, Adaptations and Mitigation of Climate Change: Scientific-Technical Analysis. Contribution of Working Group II to the Second Assessment Report of the Intergovernmental Panel on Climate Change* (Cambridge: Cambridge University Press, 1995), p. 312.

52. IPCC, *Climate Change 1995: Impacts, Adaptations and Mitigation of Climate Change: Scientific-Technical Analysis. Contribution of Working Group II to the Second Assessment Report of the Intergovernmental Panel on Climate Change* (Cambridge: Cambridge University Press, 1995), p. 368.

53. IPCC, *Climate Change 1995: Impacts, Adaptations and Mitigation of Climate Change: Scientific-Technical Analysis. Contribution of Working Group II to the Second Assessment Report of the Intergovernmental Panel on Climate Change* (Cambridge: Cambridge University Press, 1995), p. 394.

54. IPCC, *Climate Change 1995: Impacts, Adaptations and Mitigation of Climate Change: Scientific-Technical Analysis. Contribution of Working Group II to the Second Assessment Report of the Intergovernmental Panel on Climate Change* (Cambridge: Cambridge University Press, 1995), p. 452.

55. IPCC, *Climate Change 1995: Impacts, Adaptations and Mitigation of Climate Change: Scientific-Technical Analysis. Contribution of Working Group II to the Second Assessment Report of the Intergovernmental Panel on Climate Change* (Cambridge: Cambridge University Press, 1995), pp. 471–473.

56. IPCC, *Climate Change 1995: Impacts, Adaptations and Mitigation of Climate Change: Scientific-Technical Analysis. Contribution of Working Group II to the Second Assessment Report of the Intergovernmental Panel on Climate Change* (Cambridge: Cambridge University Press, 1995), p. 492.

57. IPCC, *Climate Change 1995: Impacts, Adaptations and Mitigation of Climate Change: Scientific-Technical Analysis. Contribution of Working Group II to the Second Assessment Report of the Intergovernmental Panel on Climate Change* (Cambridge: Cambridge University Press, 1995), pp. 541–551.

58. IPCC, *Climate Change 1995: Impacts, Adaptations and Mitigation of Climate Change: Scientific-Technical Analysis. Contribution of Working Group II to the Second Assessment Report of the Intergovernmental Panel on Climate Change* (Cambridge: Cambridge University Press, 1995), p. 564.

59. IPCC, *Climate Change 1995: Impacts, Adaptations and Mitigation of Climate Change: Scientific-Technical Analysis. Contribution of Working Group II to the Second Assessment Report of the Intergovernmental Panel on Climate Change* (Cambridge: Cambridge University Press, 1995), p. 217.

60. IPCC, *Climate Change 1995: Impacts, Adaptations and Mitigation of Climate Change: Scientific-Technical Analysis. Contribution of Working Group II to the Second Assessment Report of the Intergovernmental Panel on Climate Change* (Cambridge: Cambridge University Press, 1995), p. 5.

61. Rolf Lidskog and Goran Sundqvist, "Lost in Transformation: The Paris Agreement, the IPCC and the Quest for National Transformative Change," *Frontiers in Climate* 4, no. 906054 (2022): DOI 10.3389/fclim.2022.906054, p. 4.

62. IPCC, *Climate Change 2001: Impacts, Adaptation and Vulnerability. Contribution of Working Group II to the Third Assessment Report of the Intergovernmental Panel on Climate Change* (Cambridge: Cambridge University Press, 2001), Foreword.

63. IPCC, *Climate Change 2001: Impacts, Adaptation and Vulnerability. Contribution of Working Group II to the Third Assessment Report of the Intergovernmental Panel on Climate Change* (Cambridge: Cambridge University Press, 2001), p. ix.

64. IPCC, *Climate Change 2001: Impacts, Adaptation and Vulnerability. Contribution of Working Group II to the Third Assessment Report of the Intergovernmental Panel on Climate Change* (Cambridge: Cambridge University Press, 2001), p. 4.

65. IPCC, *Climate Change 2001: Impacts, Adaptation and Vulnerability. Contribution of Working Group II to the Third Assessment Report of the Intergovernmental Panel on Climate Change* (Cambridge: Cambridge University Press, 2001), p. 8.

66. IPCC, *Climate Change 2001: Impacts, Adaptation and Vulnerability. Contribution of Working Group II to the Third Assessment Report of the Intergovernmental Panel on Climate Change* (Cambridge: Cambridge University Press, 2001), p. 73.

67. IPCC, *Climate Change 2001: Impacts, Adaptation and Vulnerability. Contribution of Working Group II to the Third Assessment Report of the Intergovernmental Panel on Climate Change* (Cambridge: Cambridge University Press, 2001), p. 25.

68. IPCC, *Climate Change 2001: Impacts, Adaptation and Vulnerability. Contribution of Working Group II to the Third Assessment Report of the Intergovernmental Panel on Climate Change* (Cambridge: Cambridge University Press, 2001), p. 28.

69. IPCC, *Climate Change 2001: Impacts, Adaptation and Vulnerability. Contribution of Working Group II to the Third Assessment Report of the Intergovernmental Panel on Climate Change* (Cambridge: Cambridge University Press, 2001), p. 905.

70. IPCC, *Climate Change 2001: Impacts, Adaptation and Vulnerability. Contribution of Working Group II to the Third Assessment Report of the Intergovernmental Panel on Climate Change* (Cambridge: Cambridge University Press, 2001), p. 63, emphasis in original.

71. David Pepper, "Sustainable Development and Ecological Modernization: A Radical Homocentric Perspective," *Sustainable Development* 6, no. 1 (1998): 1–7.

72. Arran Gare, "From 'Sustainable Development' to 'Ecological Civilization': Winning the War for Survival," *Cosmos and History* 13, no. 3 (2017), p. 132.

73. IPCC, *Climate Change 2001: Impacts, Adaptation and Vulnerability. Contribution of Working Group II to the Third Assessment Report of the Intergovernmental Panel on Climate Change.*(Cambridge: Cambridge University Press, 2001), p. 63.

74. A search of the entire 1,042-page Working Group II report on "capitalism" returns no hits. The word *capital* appears on 78 pages, but as a factor of production, physical assets, or human capital, not as a comprehensive social and economic system.

75. IPCC, *Climate Change 2007: Synthesis Report. Contribution of Working Groups I, II, and III to the Fourth Assessment Report of the Intergovernmental Panel on Climate Change* (Geneva: IPCC, 2007), p. 56.

76. IPCC, *Climate Change 2007: Synthesis Report. Contribution of Working Groups I, II, and III to the Fourth Assessment Report of the Intergovernmental Panel on Climate Change* (Geneva: IPCC, 2007), p. 65.

77. IPCC, "Perspectives on Climate Change and Sustainability," in *Climate Change 2007: Impacts, Adaptation and Vulnerability. Contribution of Working Group II to the Fourth Assessment Report of the Intergovernmental Panel on Climate Change* (Cambridge: Cambridge University Press, 2007), 811–841.

78. IPCC, *Climate Change 2007: Impacts, Adaptation and Vulnerability. Contribution of Working Group II to the Fourth Assessment Report of the Intergovernmental Panel on Climate Change* (Cambridge: Cambridge University Press, 2007), p. 69.

79. IPCC, "Assessment of Adaptation Practices, Options, Constraints and Capacity," in *Climate Change 2007: Impacts, Adaptation and Vulnerability. Contribution of Working Group II to the Fourth Assessment Report of the Intergovernmental Panel on Climate Change* (Cambridge: Cambridge University Press, 2007), 717–743.

80. IPCC, "Inter-Relationships Between Adaptation and Mitigation," in *Climate Change 2007: Impacts, Adaptation and Vulnerability. Contribution of Working Group II to the Fourth Assessment Report of the Intergovernmental Panel on Climate Change* (Cambridge: Cambridge University Press, 2007), p. 765.

81. IISD, "Seventh Conference of the Parties to the UN Framework Convention on Climate Change: 20 October–10 November 2001," *Earth Negotiations Bulletin* 12, no. 189 (2001). On the conceptual difficulties attending barriers, see G. Robert Biesbroek, Judith E. M. Klostermann, Catrien J A. M. Termeer, and Pavel Kabat, "On the Nature of Barriers to Climate Change Adaptation," *Regional Environmental Change* 13, no. 5 (2013): 1119–1129; Kirstin Dow, Frans Berkhout, Benjamin L. Preston, Richard J. T. Klein, Guy Midgley, and M. Rebecca Shaw, "Limits to Adaptation," *Nature Climate Change* 3, no. 4 (2013): 305–307.

82. Annett Möhner and Richard J. T. Klein, *The Global Environment Facility: Funding for Adaptation or Adapting to Funds?* (Stockholm: Stockholm Environment Institute, 2007).

83. IISD, "UNFCCC COP-7 Highlights: Wednesday, 7 November 2001," *Earth Negotiations Bulletin* 12, no. 187 (2001), p. 2.

84. IISD, "UNFCCC COP-7 Highlights: Thursday, 8 November 2001," *Earth Negotiations Bulletin* 12, no. 188, p. 1 (2001).

85. James Ford, "Emerging Trends in Climate Change Policy: The Role of Adaptation," *International Public Policy Review* 3, no. 2 (2007): 5–16.

86. IISD, "Seventh Conference of the Parties to the UN Framework Convention on Climate Change: Monday, 29 October 2001," *Earth Negotiations Bulletin* 12, no. 179 (2001), p. 2.

87. UNFCCC, *Report of the Conference of the Parties on Its Seventh Session, Held at Marrakesh from 29 October to 10 November 2001, Addendum, Part Two, Actions Taken by the Conference of the Parties,* FCCC/CP/2001/13/Add.1, 2002.

88. Bjørn Lomborg, "The Truth About the Environment," *The Economist,* August 2, 2001: 63–65.

89. Bjørn Lomborg, *Global Warming—Are We Doing the Right Thing?* (Aarhus: University of Aarhus, 2001), p. 3.

90. UN Adaptation Committee, *25 Years of Adaptation Under the UNFCCC* (Bonn: UN Climate Change Secretariat, 2019).

91. Sam Barrett, "Local Level Climate Justice? Adaptation Finance and Vulnerability Reduction," *Global Environmental Change* 23, no. 6 (2013): 1819–1829.

92. Hans-Martin Füssel, "How Inequitable Is the Global Distribution of Responsibility, Capability, and Vulnerability to Climate Change: A Comprehensive Indicator-Based Assessment," *Global Environmental Change* 20, no. 4 (2010): 597–611.

93. Samuel Fankhauser and Guido Schmidt-Traub, "From Adaptation to Climate-Resilient Development: The Costs of Climate-Proofing the Millennium Development Goals in Africa," *Climate and Development* 3, no. 2 (2011): 94–113; Frances C. Moore, "Costing Adaptation: Revealing Tensions in the Normative Basis of Adaptation Policy in Adaptation Cost Estimates," *Science, Technology, and Human Values* 37, no. 2 (2012): 171–198.

94. UN Adaptation Fund Board, *Draft Provisional Operational Policies and Guidelines for Parties to Access Resources from the Adaptation Fund,* AFB/B/3 /8 (2008).

95. UN Adaptation Fund, "About," 2022, https://www.adaptation-fund.org.

96. Hans-Martin Füssel, "Adaptation Planning for Climate Change: Concepts, Assessment Approaches, and Key Lessons," *Sustainability Science* 2, no. 2 (2007), p. 265.

97. IISD, "Summary of the Tenth Conference of the Parties to the UN Framework Convention on Climate Change, 6–18 December 2004," *Earth Negotiations Bulletin* 12, no. 260 (2004), p. 14.

98. IISD, "Summary of the Tenth Conference of the Parties to the UN Framework Convention on Climate Change, 6–18 December 2004," *Earth Negotiations Bulletin* 12, no. 260 (2004), p. 14.

99. IISD, "Summary of the Eleventh Conference of the Parties to the UN Framework Convention on Climate Change and the First Conference of the Parties Serving as the Meeting of the Parties to the Kyoto Protocol: 28 November–10 December 2005," *Earth Negotiations Bulletin* 12, no. 291 (2005), p. 19.

100. UN Adaptation Committee, *25 Years of Adaptation Under the UNFCCC* (Bonn: UN Climate Change Secretariat, 2019), p. 13.

101. UN Adaptation Committee, *25 Years of Adaptation Under the UNFCCC* (Bonn: UN Climate Change Secretariat, 2019), p. 14.

102. Annett Möhner and Richard J. T. Klein, *The Global Environment Facility: Funding for Adaptation or Adapting to Funds?* (Stockholm: Stockholm Environment Institute, 2007).

103. IISD, "Summary of the Twelfth Conference of the Parties to the UN Framework Convention on Climate Change and the Second Meeting of the Parties to the Kyoto Protocol: 6–17 November 2006," *Earth Negotiations Bulletin* 12, no. 318 (2006), p. 1.

104. Harris, Paul G., "Collective Action on Climate Change: The Logic of Regime Failure," *Natural Resources Journal* 47, no. 1 (2007): 195–224.

105. Roger A. Pielke Jr., "The Case for Sustainable Climate Policy: Why Costs and Benefits Must Be Temporally Balanced," *University of Pennsylvania Law Review* 155, no. 6 (2007), p. 1850.

106. Roger A. Pielke Jr., Gwyn Prins, Steve Rayner, and Daniel Sarewitz, "Lifting the Taboo on Adaptation," *Nature* 445, no. 7128 (2007): 597–598.

107. Ans Kolk and Jonatan M. Pinske, "Multinationals' Political Activities on Climate Change," *Business and Society* 46, no. 2 (2007): 201–228.

108. Indur M. Goklany, "Integrated Strategies to Reduce Vulnerability and Advance Adaptation, Mitigation, and Sustainable Development," *Mitigation and Adaptation Strategies for Global Change* 12, no. 5 (2007): 755–786.

109. UNFCCC, *Report of the Conference of the Parties on Its Thirteenth Session, Held in Bali from 3 to 15 December 2007, Addendum Part Two: Action Taken by the Conference of the Parties at Its Thirteenth Session,* FCCC/CP/2007/6/Add.1, Decision 1/CP.13, 008, p. 2.

110. IISD, "Summary of the Thirteenth Conference of Parties to the UN Framework Convention on Climate Change and Third Meeting of the Parties to the Kyoto Protocol: 3–15 December 2007," *Earth Negotiations Bulletin* 12, no. 354 (2007), p. 1.

111. Delf Rothe, "Managing Climate Risks or Risking a Managerial Climate: State, Security and Governance in the International Climate Regime," *International Relations* 25, no. 3 (2011), p. 339.

112. Sven Harmeling and Christoph Bals, "Adaptation to Climate Change—Where Do We Go from Bali? Briefing Paper (Bonn: Germanwatch, 2008).

113. Clare Oh, "Adapting Is the Key," State of the Planet, Columbia Climate School, December 2, 2009, https://news.climate.columbia.edu.

114. Radoslav S. Dimitrov, "Inside UN Climate Change Negotiations: The Copenhagen Conference," *Review of Policy Research* 27, no. 6 (2010): 795–821.

115. IISD, "Summary of the Copenhagen Climate Change Conference, 7–19 December 2009," *Earth Negotiations Bulletin* 12, no. 459 (2009).

116. IISD, "Summary of the Cancun Climate Change Conference, 29 November–11 December 2010," *Earth Negotiations Bulletin* 12, no. 498 (2010), p. 1.

117. UN Adaptation Committee, *25 Years of Adaptation Under the UNFCCC* (Bonn: UN Climate Change Secretariat, 2019), p. 12.

118. UN Adaptation Committee, *25 Years of Adaptation Under the UNFCCC* (Bonn: UN Climate Change Secretariat, 2019), p. 16.

119. IISD, "Summary of the Cancun Climate Change Conference, 29 November–11 December 2010," *Earth Negotiations Bulletin* 12, no. 498 (2010), p. 9.

120. UNFCCC, *Report of the Conference of the Parties on Its Sixteenth Session, Held in Cancun from 29 November to 10 December 2010, Addendum Part Two: Action Taken by the Conference of the Parties at Its Sixteenth Session*, FCCC/CP/2010/7/Add.1, 2010, p. 2.

121. Katrina Brown, "Sustainable Adaptation: An Oxymoron?" *Climate and Development* 3, no. 1 (2011): 28–29.

122. John Barkdull and Paul G. Harris, "Emerging Responses to Global Climate Change: Ecosystem-Based Adaptation," *Global Change, Peace and Security* 31, no. 1 (2019): 19–37.

123. Dorothy Grace Guerrero, "The Limits of Capitalist Solutions to the Climate Crisis," in *The Climate Crisis: South African and Global Democratic Eco-Socialist Alternatives,* edited by Vishwas Satgar (Johannesburg: Wits University Press, 2018), p. 36.

124. John Barkdull, "The Ethics of Sustainability," *Journal of the Texas Tech Ethics Center* 4, no. 2 (2020).

125. Vito De Lucia, "The Climate Justice Movement and the Hegemonic Discourse of Technology," in *Routledge Handbook of the Climate Change Movement,* edited by Matthias Dietz and Heiko Garrelts (London: Routledge, 2013), pp. 66–83.

3

Maintaining the Global Status Quo

THIS CHAPTER CONTINUES THE EXAMINATION OF THE development of adaptation policy. It briefly surveys negotiations in the Conference of the Parties (COP) leading up to COP21 held in Paris in 2015. This includes the creation of the Warsaw International Mechanism (WIM), the first significant institutional recognition of the concept of loss and damage (L&D). In addition, the Intergovernmental Panel on Climate Change (IPCC) released its Fifth Assessment Report (AR5), encompassing the efforts of Working Groups I, II, and III. The tone and approach to adaptation changed significantly in AR5, although that was not entirely reflected in the results of international negotiations.

The narrative presented here shows that, in line with tendencies discussed in the previous chapter, adaptation has been interpreted to support the traditional development agenda, relying on economic growth and extant global institutions.[1] Despite increasing calls for transformation, the ambitions of policy remain well within the limits of the global institutional status quo. Further, the impact of adaptation discourse has been to diminish the obligation to fund adaptation on a broad scale in favor of voluntary funding provided on a small scale; assistance has come to be understood as voluntary contributions from rich countries to poor countries. In all, mainstream adaptation discourse has the effect of marginalizing *global* climate policy, framing it as a problem mainly for poor people in poor countries, and pushing implementation and funding to the local level.[2]

COP17, Durban 2011, and COP18, Doha 2012

Pursuant to the Cancun Agreements, the Subsidiary Body on Implementation (SBI) of the UN Framework Convention on Climate Change (UNFCCC) launched work on national adaptation plans, and loss and damage. The Ad Hoc Working Group on the Kyoto Protocol and the Ad Hoc Working Group on Long-Term Cooperative Action also addressed adaptation, although extending the Kyoto Protocol commitment period remained their top agenda item.

At the Durban COP17, although the dialogue had moved on somewhat, long-standing issues were still under discussion. For instance, regarding National Adaptation Programmes of Action (NAPAs), instead of seeking ways to get them written, the parties decided to "support the development of programmatic approaches for [their] implementation." However, the criteria for funding remained unclear, as parties sought "to clarify the concept of additional costs as applied to different types of adaptation projects," essential to determining how much of a project was eligible for money from the various relevant funds.[3] As before, NAPAs and the funding available to support them were to apply only to the least developed countries. Consequently, most countries were to pay for their own adaptation needs.

One notable aspect of COP17 was the attention given to ecosystem-based adaptation (EbA). In discussions of the Nairobi Work Programme (NWP), the United States urged "taking full advantage" of EbA, and the COP requested further research and information on the topic.[4] For its part, the Adaptation Fund reported that it was still at the workshop and in the planning phase. The newly established Adaptation Committee was directed to consider relevant information, provide reports, and meet requests from governments for technical support and guidance. Developing countries called for moving to implementation of "ground-based adaptation" by making funding more readily available. They hoped that this would happen due to progress made on operationalizing the Green Climate Fund and pledges by developed countries to provide $100 billion per year to it by 2020, a pledge that has not yet been fulfilled. All aspects of climate policy remained affected by debates over equity and common but differentiated responsibilities (CBDR), a debate expected to become more acute in regard to loss and damage (L&D).[5] Yet the thrust of adaptation discourse and policy remained well within established institutional lines. In all, concludes the International Institute for Sustainable Development (IISD), COP17 represented the triumph of incrementalism in a time calling for transformative action. While parties saw reason for hope in the progress on adaptation, in fact most of the activity to follow was more workshops, meetings, and reports rather than the implemented projects and programs that developing countries were demanding.

Little changed when parties met a year later in Doha for COP18. Developing countries reiterated their disappointment regarding the low level of funding for climate responses, including both mitigation and adaptation. The funds intended to provide money for adaptation projects remained short on resources, while the difficulties with allocating funding due to inadequate NAPAs and the blurry line between adaptation and development in general remained. The notion of loss and damage was controversial, as developed countries resisted the proposal to establish an institutional mechanism for L&D. Yet the following year saw the creation of just such an innovation, although its scope remained within existing commitments.

COP19: The Warsaw International Mechanism

For decades, vulnerable developing countries have sought funding to cope with the damages they have already suffered and almost certainly will experience due to climate change.[6] One outcome of this effort was the creation of the Warsaw International Mechanism in 2013. Broadly, adaptation is about making needed adjustments and transformations to infrastructure, practices, and institutions to reduce vulnerability and enhance resilience. Loss and damage was intended to add another dimension to adaptation, entering the dialogue as a call to provide compensation for the inevitable, unavoidable harms of rising global temperatures. In 1991, Vanuatu, a small island country, proposed an insurance scheme for countries likely to be affected by rising sea levels. In 2007, the Bali Action Plan explicitly mentioned L&D, albeit only once and in the context of adaptation planning. In 2010, the Cancun COP16 established a work program on L&D, aimed at better understanding the effects of weather extremes and slow-onset events on the most vulnerable countries.[7] But the radical implications of L&D have been blunted in subsequent negotiations.

Loss and damage could be interpreted as requiring payments for harms done, akin to liability in a civil lawsuit. In this context, one purpose of L&D might be to deter harmful behavior, so the financial compensation might exceed the actual damages done and it could apply retroactively.[8] Instead, L&D has become an element of adaptation, not a distinct area of action apart from mitigation and adaptation.

Accordingly, Cancun implied no new responsibilities on developed countries to accept liability and to compensate those harmed by climate change, as recent calls for climate reparations suggest. Rather, Cancun specified that the work program would focus on the effects on developing countries of extreme weather events and slow onset threats, the latter including sea level rise, increasing temperatures, ocean acidification, glacial retreat and related impacts, salinization, land and forest degradation, loss of biodiversity, and desertification.[9] Notably, the call for attention to L&D fell under the heading of "Enhanced action on adaptation." As such, the presumption is usually that the beneficiaries of adaptation (communities and countries implementing adaptation measures) will provide the bulk of the funding, while international assistance will be provided to those lacking the means for some minimal level of adaptation. L&D aims to assist those communities and countries that are most in need and facing severe impacts.

In cases in which the harms from climate change would be unavoidable and severe for a poor community or country, adaptation could entail extreme measures such as relocating entire communities. Loss and damage, then, asserts that the developed countries should provide the funding for strong measures that countries and communities must take when mitigation has failed and adaptation to avoid harms is not possible, but it does not impose punitive or compensatory damages on developed countries for the activities and policies pursued over decades that led to the problem. The demand for additional funding from developed countries to developing countries has been part of climate negotiations since they began, and L&D does not represent a break with prior assumptions about how and why such funding should be provided.

At COP19, held in Warsaw in 2013, the UNFCCC established the Warsaw International Mechanism for Loss and Damage to further policy addressing the effects of extreme events and slow onset processes on developing countries. In addition, parties created an Executive Committee to guide the WIM and provide reports on progress toward program goals. The functions of the WIM were specified:

- Enhance knowledge and understanding of comprehensive risk management approaches to address loss and damage from climate change;
- Strengthen dialogue and coordination among relevant stakeholders;

- Enhance action and support in finance, technology, and capacity building by providing technical support, information, and recommendations, and mobilizing expertise and material support.

The WIM Executive Committee encompasses five thematic expert groups, covering slow onset events, noneconomic losses, risk management, displacement, and action and support.

The WIM was to provide advice and information; it was not intended to fund the adaptation aspects of L&D, much less compensatory payments for liability for damages. Indeed, at the 2021 COP26 held in Glasgow, developing countries were disappointed that the agreements reached still did not include a financial mechanism for L&D.[10] The following year, IISD observed that four entities were charged with dealing with L&D—the WIM, the Santiago Network for technical assistance, the Fiji Clearinghouse for Risk Transfer, and the Glasgow Dialogue—none of them including funding arrangements.[11] To address this gap, COP27 created a fund for L&D and a committee to develop rules to operationalize the fund for consideration at COP28.

COP20: Lima

The main agenda item for the Lima COP was to set the stage for a comprehensive agreement to be finalized at the Paris COP21 to be held the following year. Accordingly, COP20 served as a bridge between the Durban Platform for Enhanced Action and the Paris agreement. While much of the discussion was, as before, devoted to mitigation, adaptation's prominence continued to rise at Lima, primarily at the insistence of developing countries.

The Lima conference devoted significant attention to how to define and operationalize intended nationally determined contributions (INDCs), which are each country's voluntary pledges to further climate policy. The developed countries understood INDCs to refer to mitigation, meaning countries would pledge to achieve a certain amount of reduction in their emissions and other sources of rising temperatures. Developing countries disagreed with this interpretation of the INDCs and insisted that it include an adaptation component. Developing countries' increasing insistence on the importance of adaptation affected all aspects of climate policy.

Naturally, the question of finances arose again, with the COP agreeing to urge "developed countries to channel a substantial share of public climate funds to adaptation activities."[12] The COP welcomed the work of the Standing Financial Committee on the mobilization of adaptation finance, and it called on the Standing Financial Committee to consider how to link the Adaptation Fund to other institutions. The Green Climate Fund was asked to speed up the operationalization of adaptation. The Adaptation Fund, operating under the Kyoto Protocol, urged developed countries to respond to the call for contributions, although it also noted that the Adaptation Fund Trust Fund had exceeded its funding targets, including its goal of receiving $80 million per year in contributions from developed countries. To be sure, that figure, while perhaps ambitious at the time, has been dwarfed by the current goal of achieving $100 billion per year of climate finance, with a substantial share of that expected to go to adaptation. The Subsidiary Body on Implementation focused on producing National Adaptation Plans, but it did not call for any significant change to the process. Regarding the Warsaw International Mechanism, COP statements were largely procedural and organizational, not resolving disputes over the status of L&D relative to adaptation.

The plenary of the COP adopted the Lima Call for Climate Action, which included support for the WIM, and it called for parity of treatment and provision of resources between adaptation and mitigation. However, the definition of *parity* was disputed. Developed countries sought changes in the language that would allow continued emphasis on mitigation, while developing countries wanted balance among the various elements of climate policy, in particular between mitigation and adaptation.

The growing emphasis by developing countries on adaptation was strongly evidenced at the Lima conference. Contrary to assertions that adaptation was a way for developed countries with high emissions to evade strong mitigation policy, developing countries wanted more funding for adaptation and more expression of its importance in agreements and policy statements. This included small island states and others that have made negligible contributions to the amount of greenhouse gases (GHG). In addition to advocating that INDCs include adaptation, developing countries demanded more funding for adaptation, and balanced treatment of the various components of climate policy.

In sum, the Lima COP, which was to serve as a bridge to the critical Paris COP in 2015, saw limited changes in the approach to adaptation. Unresolved questions, such as how to allocate funding for adaptation projects and the meaning of L&D, remained unresolved. The most notable aspect of the state of negotiations at this time was the rising emphasis that developing countries were placing on adaptation and their continuing insistence that adaptation funding rise to the level given to mitigation.

The IPCC Fifth Assessment Report and Transformational Change

The Intergovernmental Panel on Climate Change released its Fifth Assessment Report (AR5) in 2014, ahead of the Paris conference. AR5 expressed increasing confidence in the scientific consensus that temperatures were rising due to human activities, and that climate change would have significant effects on the natural world and human interests. It asserted that mitigation and adaptation had to work together to reduce and manage climate risks. It stated that enabling choices to limit the damage would require attention to governance, ethics, equity, and humans' wide range of orientations to risk and uncertainty. AR5 adopted sustainable development and equity as the standards by which to judge climate policies. The IPCC asserted that "insufficient adaptation responses to emerging impacts are already eroding the basis for sustainable development."[13]

Regarding GHG emissions, the IPCC reported, no doubt remained that human activity had resulted in global temperature rise, estimated at 1.1°C above the preindustrial level. Additionally, AR5 asserted that the window for holding temperature rise below 2°C was closing, and further delay would result in significant economic costs. The "multiple mitigation pathways" that could give the world a reasonable chance of limiting warming below 2°C "would require substantial emissions reductions over the next few decades and near zero emissions"[14] of GHGs by 2100. Delaying action only meant that drastic reductions would have to take place much more quickly in later decades. Otherwise, global average temperature would exceed the agreed threshold of 2°C. The trajectories observed in 2010 indicated that temperatures would increase from 2.5°C to 7.8°C when accounting for climate uncertainty. The median range would be 3.7°C to 4.8°C. These estimates became an important part of climate policy discussions.

The IPCC attributed widespread impacts to climate change, already observed in the first decade of this century. With medium to very high confidence, the IPCC reported major effects on physical, biological, and human systems, including glaciers, coastal regions, marine ecosystems, terrestrial ecosystems, agriculture, and health. Importantly, climate had affected water quantity and quality, as well as resulting in more negative effects on crop yields than positive effects. These are of particular concern as the population continues to rise and the demand for food and water with it, and it was widely recognized that societies would have to adapt to the increasingly unavoidable effects of climate change.

AR5 marked a significant change in IPCC treatment of adaptation, in particular by placing more emphasis on transformational adaptation and transformational change as the necessary responses to climate change. The report identified three major aspects of adaptation. The first, incremental adaptation, encompasses measures that do not change the basic features of a social system. Such responses limit the damage while allowing life to go on as before. The second, transformational adaptation, calls for greater change in existing patterns of life. Instead of strengthening seawalls, for example, perhaps the community moves. To be sure, the distinction between incremental and transformational adaptation is not always clear, but the IPCC asserted that the broad differences were important enough to aid in analysis of the problem.

The third major aspect of adaptation asks whether today's political and economic institutions can make the right choices regarding incremental and transformational adaptation. Because the answer increasingly appears to be no, in AR5 the IPCC suggested examining dominant development pathways and the decisionmaking framework that reinforces prevailing economic practices. This implies successful adaptation could require transformational change toward institutions that further long-term sustainability. Such questioning of prevailing institutions was not entirely absent from previous reports and official discussions, but AR5 represented a major increase in that orientation.

In line with scholarship on the topic, AR5 observed that impediments to adaptation planning and implementation arise from inadequate resources, uncoordinated governance, uncertainties, varying perceptions and values, lack of leadership and advocacy, and weak monitoring, as well as insufficient research and assessment. Political barriers also encompass cases in which short-term interests prevail when a long-term perspective is needed to guide investment, and the close relationship political leaders might have with powerful interest groups.

This listing of barriers to adaptation raises a further question: Why do they exist and why is it so hard to remove them? In every case, the prevailing institutions of the world order could be at fault. For instance, a profit-oriented economic system with a short-term perspective could stand in the way of prioritizing climate policy in general and adaptation specifically. The anarchy of the global market and the state system might hinder coordinated and integrated responses to climate change, especially in that competition and the threat of violence loom. Uncertainty would exist regardless of the institutional framework, but rational, equitable adjustments to unanticipated outcomes could be more difficult in an anarchic, competitive system. Usually, perceptions of risks and dominant values reflect the interests of the privileged and powerful sectors of global society that have the least interest in effective, equitable adaptation to climate change. Consequently, leaders and advocates who would demand alternative institutions that would better respond to the climate interests of the mass of society rather than a tiny elite are crushed and marginalized. Monitoring and research are low priorities for those who would prefer the ecological destruction of their practices not be publicized. If alternative social orders are achievable, then these impediments to adaptation are soft limits, all of which can be surmounted in a global socioeconomic order that enables responses to climate change in line with human rights and the Sustainable Development Goals. The current world order might not have that capacity, which means that it is a hard limit to adaptation as long as it persists.

To some extent, the IPCC hinted at such a sweeping indictment of what it called "current development pathways." The IPCC adopted sustainable development and equity as the basis for assessing climate policies. To be sure, it did not state that global neoliberal capitalism and equitable sustainable development are incompatible.[15] However, AR5 also asserted that employing only incremental adaptations to existing systems could risk increasing costs and losses and miss opportunities for effective adaptation. The list of barriers to effective adaptation implies that existing institutions might not be capable of more than incremental adjustments. Indeed, the very definition of *incremental* indicates as much, in that the IPCC understood incremental change to be that which preserves the basic features of a social system. Transformational adaptation, stated the IPCC, might even mean "formation of new financial structures or systems of governance."[16]

However, the IPCC was, as before, reluctant to spell out the implications of its analysis, even while calling for transformational change; it pointed to inertia in the global economic system rather than naming institutions, for instance. The new financial structures and systems of governance that the IPCC said might be required were not described. The Synthesis Report asserted that adaptation research had evolved since AR4 was released, from focusing on engineering and technological forms of adaptation to encompass ecosystem-based, institutional and social factors, and evaluation had shifted from cost-benefit analysis to include other policy and ethical orientations. But the Synthesis Report contained no mentions of capitalism and employed the term *capital* only in regard to social capital and human capital. Consequently, the IPCC did not directly confront the question of whether prevailing global institutions would enable or hinder climate policy that would contribute to sustainable development and equity. It is left to the readers to infer what new financial, social, and political systems would be required or would be optimal for achieving sustainable, equitable development in the face of climate change.

That said, the IPCC's emphasis on transformation in AR5 marked an important change in the body's conceptualization of adaptation. Looking back over a decade, AR3 tended to take existing institutions as a given, questioning the efficacy of reactive adaptation, but asserting that "climate change adaptation must consider nonclimatic stresses and be consistent with existing policy criteria, development objectives, and management structures."[17] By contrast, AR5 considered the broad context for decisionmaking on adaptation, noting that "transformational adaptation may involve decision makers questioning deep underlying principles and seeking changes in institutions, such as legal and regulatory structures underlying environmental and natural resource management, as well as in cultural values."[18]

This changed orientation was also evident in the discussions of specific areas of concern such as human security, water systems, and agriculture. The common barriers to adaptation across sectors include "inadequate governance and institutional structures at the scale of the challenge, lack of access to financial resources or relevant information for adaptation, and social and cultural norms that prevent adoption of viable adaptation options."[19] In the same vein, studies of regional adaptation showed significant shortcomings in planning and implementation due to emphasizing "short-term risk management over long-term transformative strategic planning to reduce long-term risk, which potentially increases vulnerability and therefore the costs associated with future adaptation efforts," raising the likelihood of maladaptation.[20] Although not recommended for every sector and region, transformational adaptation and transformational change were forwarded as important means for overcoming these barriers.

Still, the report's understanding of transformation remained somewhat limited. It raises the question of whether "existing institutions and systems of governance are adequate to effectively manage climate change risk." But new institutions to facilitate adaptation are conceived as agencies and similar entities created specifically to conduct research, manage climate risk, and enable implementation of adaptation measures. Noting that "new institutions do not necessarily resolve complex governance challenges," the report calls for more research on "which new institutions will be required to effectively govern adaptation."[21] Yet this recommendation remains limited in scope. Although not elaborated, it appears to suggest that improved agencies for researching, financing and, managing adaptation policy should be designed, working within the existing complex of global institutions and agencies; that is, global neoliberal capitalism and the various domestic and international agencies that manage its affairs. It was nowhere asserted that effective adaptation might require fundamental institutional transformation, encompassing class and property relations. The IPCC's recommendation that transformational adaptation might require "iterative risk management and tripleloop learning" remained within a technocratic, managerial framework, not posing any call for transformational change of social, economic, and political arrangements. Thus, the meaning of the term *transformation* remains elusive in AR5. The authors presented the possibility that transformational change might be required for equitable adaptation that supports sustainable development, but they drew back from the radical implications of that assertion.

Chapter 20 of the report on Working Group II draws together the findings to suggest how adaptation and mitigation can further sustainable development. Climate change, the authors assert, threatens sustainable development, possibly requiring transformational change, "very likely to be required for climate-resilient pathways—both transformational adaptations and transformations of social processes that make such transformational adaptations feasible."[22] This is perhaps the clearest expression of the relationship of transformational change to the implementation of adaptation: without transformations of social processes, appropriate decisions and actions will not be feasible. How does this summary chapter address the challenge of transforming social processes?

To begin, the prospects for achieving climate resilient pathways "are related fundamentally to what the world accomplishes with climate change mitigation."[23] Thus, the authors refer to a summary report by Working Group III, which focuses almost entirely on mitigation, elaborating various pathways to emissions reductions that would hold global average temperature increases below policy targets. Transformation in this context is understood to mean altering the activities that contribute to rising GHG concentrations. The policy goal the Working Group III chapter discusses is stabilizing greenhouse gases, and the main question is which transformation pathway will lead to that outcome. Achieving stable GHG concentrations will require "large-scale transformations in human societies," encompassing "the way that we produce and consume energy to how we use the land surface."[24]

The posited range of transformations leaves unaddressed the institutional transformational change that would enable stable GHG emissions at appropriately low levels. This raises significant questions about climate policy more broadly. Given that IPCC studies of both mitigation and adaptation fail to specify broad institutional transformational change, how would one know whether the same or different institutions are required to mitigate and to adapt? If the institutions differ, then will the world have to experience two transformations: one to mitigate enough for adaptation to be feasible, and another to enact appropriate adaptations to the effects of climate change that cannot be averted? How can researchers arrive at any reasonable estimation of the probability of achieving sufficient mitigation and adaptation that supports sustainable development without a clear picture of the global economic, social, and political institutions through which climate policy will be pursued? The IPCC's discussion of climate-resilient pathways does not confront these difficult questions.

Regarding adaptation per se, the IPCC recognizes the close relationship of sustainable development to resilience and vulnerability. Although positive effects of climate change on sustainable development might exist, the general view is that poorly designed responses to climate change "may derail current sustainable development policy and potentially offset already achieved gains."[25] Moreover, current development processes produce vulnerability, and sustainable development requires changes in social awareness and values that will lead to new actions and practices.

What are these new actions and practices? They include enhanced disaster risk management and anticipatory adaptation. Building adaptive capacity might mean "challenging current development models" so they better address poverty; inequality; deficits in education, health care, and safety; weak institutions; and unequal power relations. Achieving climate-resilient development requires enhanced awareness and capacity, adequate resources, and appropriate practices.[26] Practices include monitoring and planning that recognize possible needs for transformational responses, along with programs to help those most vulnerable to climate change. The global economic and political order is missing from this list. Rather, the tenor is that enhancing resilience and adaptive capacity so as to achieve sustainable development requires implementing in earnest the traditional development agenda, albeit with more emphasis than in previous IPCC reports on equity and inclusive decision-making processes.

Rising global temperatures, said the IPCC, could mandate consideration of transformational changes, in which "potentially impacted systems move to fundamentally new patterns, dynamics, and/or locations."[27] However, the new patterns and dynamics are not described; presumably, they are more equitable, inclusive decisionmaking processes leading to improved disaster risk management and anticipatory adaptation. The IPCC's concept of transformative change in AR5, then, looked a lot like relatively ambitious versions of the development agenda that has been promoted at the United Nations and in most development agencies for decades.

Attention to AR5 is important because the report was released in advance of the landmark 2015 Paris COP. The COPs over the years never implemented all the suggestions of the IPCC's Assessment Reports for climate policy, and Paris was no exception. However, the IPCC's findings influence the dialogue by framing the problem in particular ways, as well as synthesizing current thinking on important issues. AR5 represented and advanced the view that adaptation is a social, economic, and political matter, not to be comprehended in scientific and technical terms. For most scientists and policymakers, the questions pertaining to whether and why temperatures were rising had been settled. Consequently, the central concern in AR5 and at the Paris conference was what to do about it. This meant continued attention to mitigation along with rising concern for how to adapt to what would almost certainly be a significantly warmer world.

COP21: Paris

The 2015 Paris climate change conference, drawing over 36,000 participants, was a landmark in the development of all aspects of climate policy. Regarding adaptation, this conference marked a significant increase in emphasis, including adoption of the concept of a global goal to be assessed in the Global Stocktake on climate policy progress. However, the conference also revealed that long-standing issues remained in question, in particular securing sufficient financing for adaptation, balancing adaptation and mitigation, and the meaning of loss and damage. Reports appearing ahead of the Paris conference showed that adaptation was seriously underfunded compared to estimated needs. Estimates made in the previous decade suggested that developing countries would require $100 billion per year for adaptation, whereas funding via multilateral agencies amounted to about $3.9 billion.[28] Thus, the Paris conference would be addressing an area of considerable concern.

Although participants in the Paris conference placed emphasis on adaptation, the main focus was to adopt a binding agreement to reduce GHG emissions as a successor agreement to the Kyoto Protocol. Expressing the general sentiment of the assembled heads of state, high officials, and representatives of civil society organizations, President

Francois Hollande of France, head of the COP, stated that the conference would be successful if it could determine "a credible path to limit temperature rise below 2°C, or 1.5°C if possible."[29]

Even so, adaptation was not neglected, as parties noted the connections between mitigation and adaptation. By 2015, they recognized that, regardless of the extent of mitigation actions, countries would still need to adapt. In 2015, the global average temperature rise had already reached about 1°C, so some effects were already observable, and it was obvious that the concentrations of GHG in the atmosphere would continue to rise. But in preliminary negotiations, developed and developing countries differed on where to do the most, as "developed country parties called for focusing on the mitigation section of the text . . . while developing countries emphasized accelerated implementation and adaptation."[30] Commenting on the final draft, China, long an advocate of more support for adaptation, declared that the Paris agreement reflected a balance between mitigation and adaptation, but it is not clear that everyone agreed with that assessment. Meanwhile, nonstate actors deemed the entire agreement to be inadequate and claims of significant progress to be delusional.

At Paris, L&D remained tied to adaptation; one group was charged with negotiating both topics. Having ensured that L&D would not entail acceptance of liability for harms due to climate change, developed countries could tolerate inclusion of L&D in the text as a form of reactive adaptation. Indeed, the United States asserted that it would support efforts to avert and minimize loss and damage from climate change, but only on a basis that did not involve liability and compensation, and a paragraph in the agreement makes that clear.[31]

Article 7 of the Paris agreement deals directly with adaptation, encompassing several principles and proposed actions. First, the parties agreed to establish a global goal on adaptation. This was intended to enhance adaptive capacity, strengthen resilience, and reduce vulnerability. Adaptation was expected to contribute to sustainable development and be sufficient to respond to the temperature rises anticipated in the mitigation elements of the negotiations (2°C, 1.5°C if possible). Further, the Paris agreement deemed adaptation a global challenge, not only a matter for local or national responses. The parties to the agreement recognized that adaptation needs were significant, that mitigation could reduce adaptation needs, and that adaptation should be "country-driven, gender-responsive, participatory and fully transparent."[32] It should give particular attention to vulnerable developing countries. The agreement called for enhanced cooperation on sharing information, reforming institutions, and developing good scientific knowledge to support adaptation. All parties were mandated to undertake adaptation planning and implementation. Parties were also asked to provide ample information on their plans, practices, and experiences with adaptation, and a public registry of adaptation actions was to be created. The Global Stocktake was expected to recognize the actions of developing countries, improve communication on adaptation experience, evaluate adaptation, and assess the adequacy of support provided for adaptation.[33]

Article 8 of the agreement addressed L&D, defining it in such a way that it fit readily into the adaptation framework. The specific focus of L&D was preparing the least developed countries for the increasing frequency and severity of extreme weather events, and the impacts of slow onset events. Regarding the first element, the Paris agreement called for early warning systems and emergency preparedness, and risk insurance. The agreement calls for international cooperation to address slow onset events, unavoidable and permanent loss and damage, and enhancing resilience. Consequently, L&D in the Paris agreement was framed as a form of mainly reactive adaptation to the increasingly certain effects of rising temperatures.

Articles on finance, technology, capacity building, and others include exhortation to balance support for mitigation and adaptation, with particular concern for the needs of developing countries. The COP requested UN entities to recognize and assess adaptation efforts in developing countries. It also asked for greater cooperation among parties, and further support for producing National Adaptation Plans.

The lack of concrete progress on adaptation is reflected in the IISD's obscure characterization of the COP decision, which stated that the parties "decide on the process to assess progress made in the process to formulate and implement national adaptation plans (NAPs)."[34] Similarly, the COP requested the Standing Committee on Finance and other relevant institutions "to develop methodologies and make recommendations for consideration and adoption by CMA 1 on taking the necessary steps to facilitate the mobilization of support for adaptation in developing countries."[35] Such jumbled phrases calling, in essence, for a commitment to make a commitment to do something indicate that the parties could not actually agree on what was to be done to comply with the broad goals of enhancing resilience and reducing vulnerability. This meant especially not making any firm commitments to fund adaptation, certainly not at the level anticipated in the pledge by the developed countries to provide $100 billion per year for climate measures.

The rising awareness of the threat of maladaptation has added to the difficulties of formulating clear adaptation policy. Designing appropriate adaptation is challenging in the face of uncertainties about the future impacts of climate change, as well as variations across regions and sectors. Failing on these dimensions can render an adaptation measure inappropriate or even harmful. Accordingly, maladaptation is defined as efforts to adapt that result in higher costs and vulnerability in the future or to other groups and sectors.[36] It refers to measures taken to respond to climate change that "undermine capacities or opportunities for present and future adaptation."[37] For example, installing air conditioning to cope with higher temperatures could result in more GHG emissions as demand for electricity rises and is met with fossil fuel–fired power plants. The IPCC noted that maladaptation can result from poor planning and implementation, short-term thinking, and failure to anticipate the consequences of climate change.[38] The concept of maladaptation could imply the need for sweeping transformational change in social, economic, and political institutions at all levels. In the broader adaptation framework of AR5, any adaptation that might thwart the process of transformational change could be seen as maladaptation. That is, ensuring the perpetuation of institutions that would be unable to cope with the adaptation needs of a much warmer world would undermine the prospects of future adaptation. If, as the IPCC and numerous scholars have suggested, prevailing institutions are barriers to adaptation that is equitable and that supports sustainable development, then any adaptation that supports the institutional status quo is maladaptation. Only those measures that are compatible with or promote transformational change of the global order are true adaptation. Although AR5 discussed maladaptation in depth, parties at the Paris conference did not devote significant attention to it.

Ecosystem-based adaptation drew attention at the Paris meeting and has become more prominent in climate adaptation discussions since.[39] The idea behind EbA is that ecosystem services can be harnessed to reduce vulnerability and increase resilience of threatened natural and human systems. For instance, replanting forests can help reduce the impact of heavy rains, and encouraging the growth of mangroves can protect coastlines. Some EbA projects have been implemented, and some existing projects have been reframed as examples of EbA. Advocates of EbA hoped that it would avoid the problems of maladaptation attending technological fixes, and they called for inclusive decision-making that would draw in particular on Indigenous knowledge. However attractive that EbA might be, it would not likely have much effect if global temperatures rise beyond the globally agreed thresholds of

1.5°C or 2.0°C. Beyond that, the effects of global warming on the ecosystem would likely exceed natural limits for sustaining many EbA projects, so EbA must rely on effective mitigation policy; whether that prerequisite will be present is very much in doubt. Consequently, EbA highlights the close relationship between mitigation and adaptation, as well as indicating the limits to adaptation.

One of the notable results of the Paris conference was adoption of the idea of a Global Goal on Adaptation (GGA). The GGA encompasses the familiar elements of adaptation policy; namely, adaptive capacity, resilience, and vulnerability. The GGA reflects developing countries' insistence on raising the profile of adaptation as climate impacts were being felt and were sure to become more severe in the future. However, identifying the components of the goal and how to measure progress toward it has been challenging. Unlike mitigation, which can be measured in terms of atmospheric concentrations, emissions per year, and various other quantitative indicators, adaptation is "context-specific, and metrics of progress are difficult to aggregate globally."[40] The UNFCCC addressed these difficulties at COP26 in 2021, establishing the two-year Glasgow–Sharm el-Sheikh Work Programme on the Global Goal on Adaptation. The UN bodies charged with conducting the work were asked to craft a global goal and measures that could assess enhanced adaptation.[41] Thus, six years after Paris, basic terms for assessing the global goal remained undefined. A report on the progress of the work program suggests that clarity on key terms remains elusive.[42]

Assessing Paris

The political leaders at the Paris conference declared the outcome to be a significant achievement: ambitious, binding, universal, heralding a new era of multilateral, demonstrating global cooperation, representing solidarity with the poor, and being a boon to diplomacy and technology.[43] By contrast, critics focusing on the mitigation aspect of the agreement declared it a sham, a fraud, worthless words, inconsistent with the science, and, at best, little more than one step on a long road to an effective response to climate change. Meanwhile, outside the meeting halls, the impacts of rising global temperatures were manifesting, as outlined in the IPCC's reports and in daily news stories. In that world, the consensus to hold temperature rise below 2.0°C or, writes Brian Tokar, even to 1.5°C is "far too little, far too late." Whether embracing an optimistic or pessimistic view of the mitigation future, nearly all participants in climate policy and politics recognized that adaptation would play a significant part. Indeed, one critical observer claimed that, in light of feeble efforts to stem rising temperatures, "the real plan going forward is one of *adaptation* on a continuously heating planet."[44]

The heads of state and leading officials who negotiated the Paris agreement would certainly not agree that the "real plan" is only to adapt. However, all parties and observers agreed by 2015 that adaptation would be an important part of the response. Further, in line with IPCC reports, advocacy, and scholarly research, the notion of transformational rather than incremental responses had come to the fore in international negotiations.

The purpose of this review of the historical record has been to ask whether climate policy heralds significant change in the world order or has itself been enfolded in the prevailing institutions and practices of global politics. If adaptation's transformational potential were to be manifested, it would have been evident in the processes surrounding the landmark Paris conference. By then, adaptation had been part of the policy dialogue for over thirty-five years, beginning in the meetings of scientists that resulted in the creation of the IPCC.

Although adaptation discussions might have suggested the need for radical change at certain points, in practice it has invariably been interpreted in ways compatible with existing arrangements. This is not because adaptation is the easier path compared to mitigation, or that adaptation is emphasized when actors refuse to accept the radical changes needed to mitigate enough to hold temperatures below the danger thresholds. Adaption, instead, also carries high costs and radical implications, perhaps mandating sweeping institutional change. Indeed, adapting to a 3°C or 4°C world might prove a greater challenge than avoiding that outcome to begin with, so adaptation is hardly the easy way out. Instead, the radical message of both mitigation and adaptation has been tamed.

The official version of climate policy remains well within the existing institutional framework. Mitigation, we are told, calls for carbon markets and the willing participation of fossil fuel multinationals, as well as eco-friendly individual lifestyle choices. Green capitalism with renewables is the rosy future. Adaptation, rather than requiring transformational change of institutions so the right choices and proper allocation of resources can be made, is best managed through development-oriented agencies. Fairly modest amounts of financial support for the least developed countries can take care of the worst impacts while the rest of the world carries on with business as usual.

This orientation at Paris showed up most starkly in the treatment of L&D. The developed countries had sought for years to ensure that L&D did not imply liability and compensation for the harms that had befallen developing countries, nor those to come. Instead, adaptation assistance under the rubric of L&D was to be seen as a form of foreign aid closely tied to the development agenda. This move rendered L&D a form of adaptation focused on a particular set of problems facing a limited number of poor countries. For the rest of the world, most adaptation was to be managed as a local or national problem, mostly paid for with local and national funds. Thus, L&D, rather than constituting a wide-ranging demand for redistribution and accountability, joined other adaptation measures as a sign of the generosity of the wealthy nations toward the disadvantaged ones. This fit well with the effort on the mitigation side to do away with the principle of common but differentiated responsibilities. The message in both instances was that we are all responsible and all must contribute to the solution, although some might need help to get there.

In this context, it is important to note a shift in attitude by the developing countries. When climate negotiations began, highly vulnerable countries, such as the small island states, emphasized mitigation. Their hope was that the problem could be averted and the severe impacts they might experience under business-as-usual conditions would not happen. By the early part of this century, hopes had faded and developing countries expected that they would have to adapt, incurring significant costs to do so. Thus, their demands that adaptation be placed on the same level as mitigation became more insistent, affecting the character of negotiations and agreements, if not the actual allocation of resources. Contrary to the view that the main advocates of adaptation were oil companies and large oil-producing nations, developing countries took the lead at Paris and well before.

The Paris meeting displayed the status of adaptation policy. Parties mostly agreed that adaptation should have equal priority to mitigation, although funding still lagged and has never been sufficient. Adaptation became increasingly tied to sustainable development. The criterion for good adaptation was, in essence, that it would contribute to sustainable development as defined and understood in such organizations as the United Nations. Transformational adaptation became a notable part of the adaptation dialogue, albeit with widely different understandings of what it entailed. Transformational change received rhetorical acknowledgment, but little in climate agreements and policy suggested that the world was ready to make significant changes to existing practices.

This draws attention to the question of what transformational change could mean and what it could suggest for a different world order.

After Paris

To date, seven COPs have taken place since Paris COP21: Marrakech, Morocco; Bonn, Germany; Katowice, Poland; Madrid, Spain; Glasgow, Scotland; Sharm el-Sheik, Egypt; and Dubai, United Arab Emirates. The 2020 COP was canceled due to the Covid-19 pandemic. COP21 adopted the Paris Agreement Work Programme (PAWP) to "develop the Agreement's operational details" working through the newly created Ad Hoc Working Group for the Paris Agreement (APA), as well as the Subsidiary Body for Implementation, and the Subsidiary Body for Scientific and Technological Advice (SBSTA).[45] Subsequent work aimed mainly at meeting mandates outlined in the PAWP. This section summarizes developments in adaptation policy during those years, 2016 to present. In general, whereas one might have expected resolution of long-standing issues to be closer after more than two decades of negotiations, the record indicates no diminution of contention and disagreement, including on how to manage adaptation.

Similar to earlier years, international negotiations after Paris retained a strong concern with financing adaptation to climate change. Developed countries continued to insist that the main source of adaptation support should be public funds from developed countries to finance climate action in developing countries. But developed countries wanted private sector expenditures to count and greater reliance on market mechanisms; indeed, they wanted to drop the subject of long-term climate finance from high-level discussions.[46] Unfortunately, the shortfall in funding is huge. The report of the 2023 COP asserted that, for developing countries, adaptation funding alone should be $215 billion to $387 billion annually until 2030, considerably more than the promised, but never delivered, $100 billion per year pledged for all climate finance. This pales next to the estimated cost of shifting to clean energy so as to achieve net zero, $4.3 trillion per year until 2030, rising to $5 trillion per year until 2050.[47] Without that immense investment, incremental adaptation might be largely ineffective, and the cost of adaptation could rise commensurately. To be sure, $5 trillion represents perhaps 5 percent of gross world product and so might appear quite manageable, but not even defense spending consumes that much of world output. It would be remarkable if the international community could commit to such a large outlay while also attending to other priorities. Moreover, as one might expect, wealthier countries are averse to sending such large sums to developing countries to pay for their adaptation needs. Consequently, before and after Paris, paying for measures to ameliorate the harms from climate change has remained contentious and insufficient.

Loss and damage under the Warsaw International Mechanism was discussed primarily as a matter of funding, as particularly vulnerable developing and poor countries wanted access to money that is additional to development assistance and other adaptation funding. Developed countries resisted, noting that existing funding entities were already supporting L&D, and that developing countries should make efficient use of existing resources. At the 2019 COP25, parties limited actions on L&D to procedural steps without addressing the substance of disagreements over the purpose of and funding for L&D. A fund for L&D was established at COP27 in 2022, with the rules to be worked out the following year. This instance indicates the tenor of debates over financing after Paris; that is, things were much the same as before, and the level of financial commitment for adaptation, including L&D, remains well below estimated needs.

The concern for sufficient adaptation finance reflected continuing questions about whether adaptation would be balanced with mitigation. Yet a gap remains even after so many conferences and meetings at which balance has been urged. The Dubai COP28 emphasized the "urgent need to scale up adaptation finance," calling for doubling of adaptation funding "to achieve a balance between mitigation and adaptation in the provision of scaled-up financial resources." Yet even assessing the level of effort has proven difficult. "In 2021," notes the IISD, "developed countries promised to double adaptation finance levels above 2019 levels," but there was no agreement on the level in 2019, estimates ranging from $7.1 billion to $19.2 billion.[48] Obviously, even doubling the highest estimate would leave adaptation funding far short of the promised balanced treatment of adaptation and mitigation in a promised $100 billion annual outlay.

As noted, the Paris COP had drawn attention to adaptation by requiring a Global Stocktake on both mitigation and adaptation. Subsequent work attempted to gather the information needed to fulfill that mandate. The Global Stocktake for Adaptation was to be delivered in 2023. Instead, plans for producing the report were outlined at the 2023 Dubai COP. To produce the Global Stocktake requires that countries submit sufficient information to assess whether they are meeting mitigation pledges and implementing measures to adapt to climate change compatible with climate resilient development and the Sustainable Development Goals. Without these country reports, it is difficult to know how well the collective effort is performing. However, obtaining such information has not proven easy. For instance, countries at the 2019 COP25 could not agree on a schedule for submitting the reports. Some countries wanted a ten-year cycle, while others wanted reports submitted (or at least updated) every five years. On the longer schedule, two Global Stocktakes would have taken place before some countries had provided information in nationally determined contributions.[49] In 2023, conference leaders urged submission of timely information and adequate funding for production of National Adaptation Plans and other vital documents.

Related to the Global Stocktake, the UNFCCC had also decided to establish a Global Goal on Adaptation at the Paris conference. This is a difficult enterprise. Mitigation goals can be quantified in terms of atmospheric concentrations of GHG, annual emissions, and the like. Adaptation, by contrast, offers no such measures. Consequently, after years of discussions of the global goal, parties at the most recent COP, held in Dubai, only then agreed to a framework for assessment. The framework "aims to guide the implementation of the goal and, among other things, establishes impact, vulnerability, and risk assessment (by 2030), multi-hazard early warning systems (by 2027), climate information services for risk reduction and systematic observation (by 2027), and country-driven, gender-responsive, participatory, and transparent national adaptation plans (by 2030)."[50] Implementation of elements of this framework has supposedly been under way for many years, yet the most recent conference still treated them as future actions, suggesting that progress on these fronts is halting, at best. Clearly, agreeing on a framework for policy at such a late date might also signal lack of actions on the ground to adapt to a warmer world.

The principle of common but differentiated responsibilities, taking account of different levels of capability, was intended to guide implementation of the Paris agreement, but parties were still debating what that meant and whether it even remained relevant years after Paris. At COP28 in 2023, developing and developed countries continued to differ on the relevance and applicability of CBDR. Developed countries suggested that the Paris agreement stood apart from the UNFCCC, and so perhaps CBDR did not hold for Paris or for the Paris-mandated Global Stocktake. Developing countries argued the contrary, that Paris was governed by the UNFCCC and, thus, so was the Global Stocktake. This debate arose in regard to adaptation in a discussion of the Global Goal on Adaptation, where "parties diverged" on whether to include principles of equity and CBDR in decision documents.[51] In short, decades of debate on CBDR had not resolved basic issues in its application.

The Paris agreement adopted the goal outlined in AR6 of achieving climate-resilient development (CRD). This would mean integrating mitigation, adaptation, and sustainable development so that each goal would support the others, and the outcome would be holding global temperature rise below dangerous levels, preparing for unavoidable and harmful impacts of global warming, and enhancing development goals of poverty reduction and better human well-being. Presumably, if any one of the three components of CRD falls short, achievement of the other two would be hindered and perhaps prevented. Reviewing the discourse on adaptation since Paris suggests that adaptation policy is running in place, with the same issues arising and the same calls for action made year after year while minimal action on the ground occurs in the face of record heat. After reviewing the results of the 2023 Dubai climate conference, the IISD concluded, "There are gaps galore on emissions, adaptation, and finance."[52]

The most recent UN Environment Programme (UNEP) Adaptation Gap Report (AGR), tellingly subtitled "Underfinanced and Underprepared," reviews the progress toward effective adaptation and the much greater signs of inadequate preparation for a significantly warmer world.[53] The AGR estimates the annual adaptation finance gap as ranging from $194 billion to $366 billion, based on 2021 adaptation public finance flows of $21 billion per year.[54] Indeed, funding had decreased by 15 percent from 2020 to 2021, leaving a finance gap ten to eighteen times larger than annual public finance flows. The pledge to double adaptation funding, if met, would reduce that gap by only 5 to 10 percent. The gap becomes immensely larger when the basis for estimating financial needs is information provided by developing countries; UNEP's analysis suggests adaptation finance needs could be as high as $975 billion annually.[55] The AGR observes that an important way to close the gap would be increased domestic expenditures on adaptation, but the developing countries require assistance because addressing poverty and development needs absorbs their resources. Moreover, shifting the burden to them runs directly against the emerging support for L&D, which assumes the countries that have caused the damage should pay for the remedies. We can expect a widening finance gap, commensurate with a worsening climate crisis and more loss and damage.

Reviewing the overall status of adaptation policy, the AGR finds some evidence of progress in planning, but less to suggest that sufficient adaptation is taking place on the ground to meet the challenge. Recognition of L&D, adoption of a framework for the global goal on adaptation, the Global Stocktake, and the steady rise in the number of countries with some form of planning instrument in place are hopeful signs, asserts the AGR. However, concedes the AGR, the components of the global goal framework were "still under discussion," and identifying appropriate targets and metrics for assessing progress on adaptation has proven challenging. Indeed, the work of bodies charged with producing information "on the adequacy and effectiveness of adaptation and support is not yet advanced enough to benefit the completion of the first global stocktake."[56] The report states that 85 percent of countries have adopted at least one national-level planning instrument, such as a policy, strategy, plan, and regulations, and half of the remaining 15 percent have a process to produce a planning instrument for adaptation under way. Further, 46 of 155 UNFCCC-defined developing countries had produced NAPs, helpful to access loans and grants from international funding agencies. However, translating planning into action has proven difficult. Too, the assessment step in the policy process has been fulfilled by only a handful of countries, with no small island developing states having produced an evaluation report, despite their status as the most vulnerable to climate change. Thus, UNEP reports increased ambition for plans and commitments, "but there also remains an implementation gap, in that plans are implemented inadequately, unevenly, and incrementally,"[57] and comprehensive, adequate reporting to facilitate monitoring and assessment of adaptation projects and actions is scarce.

In 2022, the international agencies charged with funding adaptation provided $559 million to new projects, 10 percent higher than the average of the previous five years. Thirty-five countries listed 1,100 projects in their adaptation communications, but details were provided for only about half, and information on outcomes was offered for only 6 percent of the 670 for which details were available. Relying on a limited set of cases (reports from 35 countries on 670 actions), the AGR notes that half of reported actions were "preparatory," directed toward "building the institutional and knowledge base for more substantive adaptation actions."[58] This situation persists over a decade after the UNFCCC called for submission of National Adaptation Plans, and with widespread recognition of the growing importance of anticipatory adaptation. Projects targeting multiple sectors, as well as those targeting agriculture and biodiversity, accounted for 69 percent of all actions for which information was available. Of those reports identifying vulnerable groups, the largest share directed support to farmers (46 percent), with women second (14 percent). Implementation of projects thus did not emphasize vulnerable groups defined by gender, age, ethnicity, and the like. The barriers to enhanced adaptation are familiar: lack of information, poor coordination, inattention to adaptation in development planning, and inadequate or inaccessible financial resources.[59]

Significantly, UNEP observes that the costs of incremental adaptation determine the size of the adaptation gap estimated in the AGR. If transformative adaptation or transformational change are required, as many analysts believe, then "adaptation will involve very different costs,"[60] which are likely to be much higher. Consequently, a close examination of transformation is in order, the topic of Chapter 4.

Considering the changes in orientations to climate policy over decades, there has been a significant shift. As we have seen, adaptation was part of the early discussions of climate change, when scientists did not expect much policy response and thought that humans would adapt to changing conditions without significant government direction. However, as fears about the magnitude of impacts rose, that view quickly gave way to the focus on reducing greenhouse gases in time to avert severe impacts on society and human well-being, as well as the ecosystem. Years of negotiations followed on how to mitigate global warming, primarily through a transition from fossil fuels to renewables, along with other measures. But although the projected temperature in 2100 is about 1°C lower after Paris than it would have been without current pledges, it remains that mitigation policy did not achieve enough reduction in projected temperature rise to carry on as usual. Accordingly, attention shifted to adaptation, including proactive measures that would protect society from the inevitable threats arising from a much warmer climate. Yet as UNEP has reported for years, adaptation has fallen well short of reducing vulnerability and increasing resilience enough to avert serious harms, and this is already evident, well before global average temperature peaks. It was always expected that adaptation would include addressing harms that could not be avoided, but it was thought that the scope of such challenges would be manageable. Now, it appears that global average temperature will surpass thresholds for dangerous climate change, severe damages will be widespread and costly, preparations for that world have already fallen behind needs, and the situation will worsen. Thus, the focus recently has been on addressing loss and damage, which refers to unavoidable, severe harms due to climate change. In short, then, the changing orientation of climate policy reflects a record of policy failure, leading now to consideration of how to salvage something from the anticipated wreckage.[61]

Notes

1. Jessica Ayers, and David Dodman, "Climate Change Adaptation and Development, pt. I: The State of the Debate," *Progress in Development Studies* 10, no. 2 (2010): 161–168.

2. In domestic politics, countries vary regarding the emphasis placed on adaptation. The global framing of the problem does not mean that a given developed country will ignore or downplay adaptation. To the contrary, most developed countries have their own plans and programs for addressing the effects of climate change.

3. IISD (International Institute for Sustainable Development), "Summary of the Durban Climate Change Conference 28 November–11 December 2011," *Earth Negotiations Bulletin* 12, no. 534 (2011), p. 7.

4. IISD, "Summary of the Durban Climate Change Conference 28 November–11 December 2011," *Earth Negotiations Bulletin* 12, no. 534 (2011), p. 9.

5. IISD, "Summary of the Durban Climate Change Conference 28 November–11 December 2011," *Earth Negotiations Bulletin* 12, no. 534 (2011), p. 30.

6. Lisa Schultheiß, Zoha Shawoo, Inès Bakhtaoui, Lina Ahmed, Courtney Lindsay, and Arunima Sircar, *Operationalising the Loss and Damage Fund: Learning from the Funding Mosaic* (Bonn: Germanwatch e.V., 2023).

7. IISD, "Summary of the Cancun Climate Change Conference," *Earth Negotiations Bulletin* 12, no. 498 (2010), p. 17.

8. Daniel A. Farber, "Adapting to Climate Change: Who Should Pay," *Journal of Land Use and Environmental Law* 23, no. 1 (2007), pp. 19–22.

9. UNFCCC (UN Framework Convention on Climate Change), *Report of the Conference of the Parties on Its Sixteenth Session, Held in Cancun from 29 November to 10 December 2010, Addendum Part Two: Action Taken by the Conference of the Parties at Its Sixteenth Session,* FCCC/CP/2010/7/Add.1, 2010, p. 6.

10. IISD, "Glasgow Climate Change Conference: 31 October–13 November 2021," *Earth Negotiations Bulletin* 12, no. 793 (2021): pp. 2 and 16.

11. IISD, "Summary of the Sharm el-Sheikh Climate Change Conference: 6–20 November 2022," *Earth Negotiations Bulletin* 12, no. 818 (2022), p. 9.

12. IISD, "Summary of the Lima Climate Change Conference: 1–14 December 2014," *Earth Negotiations Bulletin* 12, no. 619 (2014), p. 5.

13. IPCC (Intergovernmental Panel on Climate Change), *Climate Change 2014: Synthesis Report. Summary for Policymakers* (Geneva: IPCC, 2014), p. 15. Similarly, scholars found that adaptation decisions were "mostly incremental and focused on proximate causes," with little attention to transformational change; see Russ M. Wise, Ioan Fazey, Mark Stafford Smith, Sarah E. Park, Hallie C. Eakin, Emma R. M. Archer Van Garderen, and Bruce M. Campbell, "Reconceptualising Adaptation to Climate Change as Part of Pathways of Change and Response," *Global Environmental Change* 28 (2014): 325–336.

14. IPCC, *Climate Change 2014: Synthesis Report. Summary for Policymakers* (Geneva: IPCC, 2014), p. 20.

15. IPCC, *Climate Change 2014: Synthesis Report. Contribution of Working Groups I, II and III to the Fifth Assessment Report of the Intergovernmental Panel on Climate Change* (Geneva: IPCC, 2014), p. 76.

16. IPCC, *Climate Change 2014: Synthesis Report. Contribution of Working Groups I, II and III to the Fifth Assessment Report of the Intergovernmental Panel on Climate Change* (Geneva: IPCC, 2014), p. 80.

17. IPCC, *Climate Change 2001: Impacts, Adaptation, and Vulnerability. Contribution of Working Group II to the Third Assessment Report of the Intergovernmental Panel on Climate Change* (Cambridge: Cambridge University Press, 2001), p. 62.

18. IPCC, "Foundations for Decision-Making," in *Climate Change 2014: Impacts, Adaptation and Vulnerability. Contribution of Working Group II to the Sixth Assessment Report of the Intergovernmental Panel on Climate Change* (Cambridge: Cambridge University Press, 2014), p. 218.

19. IPCC, "Adaptation Opportunities, Constraints, and Limits," in *Climate Change 2014: Impacts, Adaptation and Vulnerability. Contribution of Working Group II to the Sixth Assessment Report of the Intergovernmental Panel on Climate Change* (Cambridge: Cambridge University Press, 2014), p. 922.

20. IPCC, "Adaptation Opportunities, Constraints, and Limits," in *Climate Change 2014: Impacts, Adaptation and Vulnerability. Contribution of Working Group II to the Sixth Assessment Report of the Intergovernmental Panel on Climate Change* (Cambridge: Cambridge University Press, 2014), p. 924.

21. IPCC, "Adaptation Opportunities, Constraints, and Limits," in *Climate Change 2014: Impacts, Adaptation and Vulnerability. Contribution of Working Group II to the Sixth Assessment Report of the Intergovernmental Panel on Climate Change* (Cambridge: Cambridge University Press, 2014), p. 927.

22. IPCC, "Climate-Resilient Pathways: Adaptation, Mitigation, and Sustainable Development," in *Climate Change 2014: Impacts, Adaptation and Vulnerability. Part A: Global and Sectoral Aspects. Contribution of Working Group II to the Fifth Assessment Report of the Intergovernmental Panel on Climate Change* (Cambridge: Cambridge University Press, 2014), p. 1106.

23. IPCC, "Climate-Resilient Pathways: Adaptation, Mitigation, and Sustainable Development," in *Climate Change 2014: Impacts, Adaptation, and Vulnerability. Part A: Global and Sectoral Aspects. Contribution of Working Group II to the Fifth Assessment Report of the Intergovernmental Panel on Climate Change* (Cambridge: Cambridge University Press, 2014), p. 1104.

24. IPCC, "Assessing Transformation Pathways," in *Climate Change 2014: Mitigation of Climate Change. Contribution of Working Group III to the Fifth Assessment Report of the Intergovernmental Panel on Climate Change* (Cambridge: Cambridge University Press, 2014), p. 490.

25. IPCC, "Climate-Resilient Pathways: Adaptation, Mitigation, and Sustainable Development," in *Climate Change 2014: Impacts, Adaptation and Vulnerability. Part A: Global and Sectoral Aspects. Contribution of Working Group II to the Fifth Assessment Report of the Intergovernmental Panel on Climate Change* (Cambridge: Cambridge University Press, 2014), p. 1109.

26. IPCC, "Climate-Resilient Pathways: Adaptation, Mitigation, and Sustainable Development," in *Climate Change 2014: Impacts, Adaptation, and Vulnerability. Part A: Global and Sectoral Aspects. Contribution of Working Group II to the Fifth Assessment Report of the Intergovernmental Panel on Climate Change* (Cambridge: Cambridge University Press, 2014), p. 1113.

27. IPCC, "Climate-Resilient Pathways: Adaptation, Mitigation, and Sustainable Development," in *Climate Change 2014: Impacts, Adaptation and Vulnerability. Part A: Global and Sectoral Aspects. Contribution of Working Group II to the Fifth Assessment Report of the Intergovernmental Panel on Climate Change* (Cambridge: Cambridge University Press, 2014), p. 1116.

28. UNEP (UN Environment Programme), "Emission and Adaptation Gaps: Can We Bridge the Cracks in Climate Policy?" UNEP Global Environmental Alert Service, 2014.

29. IISD (2015), "Summary of the Paris Climate Change Conference: 29 November–13 December 2015," *Earth Negotiations Bulletin* 12 (663), p. 3; IISD (2015) "Paris Highlights: Monday, 7 December 2015," *Earth Negotiations Bulletin* 12 (659).

30. IISD (2015), "Summary," p. 8.

31. Brian Tokar, "The Paris Climate Agreement: Hope or Hype?" ZNET, December 30, 2015, https://znetwork.org.

32. IISD, "Summary of the Paris Climate Change Conference: 29 November–13 December 2015," *Earth Negotiations Bulletin* 12, no. 663 (2015), p. 26.

33. IISD, "Summary of the Paris Climate Change Conference: 29 November–13 December 2015," *Earth Negotiations Bulletin* 12, no. 663 (2015), p. 16.

34. IISD, "Summary of the Paris Climate Change Conference: 29 November–13 December 2015," *Earth Negotiations Bulletin* 12, no. 663 (2015), p. 1.

35. IISD, "Summary of the Paris Climate Change Conference: 29 November–13 December 2015," *Earth Negotiations Bulletin* 12, no. 663 (2015), p. 20.

36. Benzie Magnus, "Social Justice and Adaptation in the UK," *Ecology and Society* 19(1) (2014): 39.

37. Alexandre Magnan, "Avoiding Maladaptation to Climate Change: Towards Guiding Principles," *S.A.P.I.E.N.S.* 7, no. 1 (2014), p. 3.

38. IPCC, *Climate Change 2014: Synthesis Report. Contributions of Working Groups I, II and III to the Fifth Assessment Report of the Intergovernmental Panel on Climate Change* (Geneva: IPCC, 2014), p. 20.

39. John Barkdull and Paul G. Harris, "Emerging Responses to Global Climate Change: Ecosystem-Based Adaptation," *Global Change, Peace and Security* 31, no. 2 (2019): 19–37.

40. Kiyomi de Zoysa, Tamara Coger, and Nisha Krishnan, "Can the Global Goal on Adaptation Be Locally Led?" (Washington, DC: World Resources Institute, July 22, 2022).

41. UNFCCC, *Glasgow-Sharm el-Sheikh Work Programme on the Global Goal on Adaptation,* n.d., https://unfccc.int/topics.

42. UNFCCC, *Workshops Under the Glasgow–Sharm el-Sheikh Work Programme on the Global Goal on Adaptation,* Report by the Secretariat, FCCC /SB/2022/INF.2, 2022.

43. Brian Tokar, "The Paris Climate Agreement: Hope or Hype?" ZNET, December 30, 2015, https://znetwork.org.

44. Carl Boggs, "The Grand Illusion," ZNET, November 30, 2019, https://znetwork.org, emphasis in original.

45. IISD, "Summary Report, 2–15 December 2018: Katowice Climate Change Conference—December 2018," *Earth Negotiations Bulletin* 12, no. 747 (2018), p. 1.

46. IISD, "Summary of the Chile/Madrid Climate Change Conference: 2–15 December 2019," *Earth Negotiations Bulletin* 12, no. 775 (2019), p. 18.

47. IISD, "Summary of the 2023 Dubai Climate Change Conference: 30 November–13 December 2023," *Earth Negotiations Bulletin* 12, no. 842 (2023), p. 7.

48. IISD, "Summary of the 2023 Dubai Climate Change Conference: 30 November–13 December 2023," *Earth Negotiations Bulletin* 12, no. 842 (2023), p. 29.

49. IISD, "Summary of the Chile/Madrid Climate Change Conference: 2-15 December 2019," *Earth Negotiations Bulletin* 12, no. 775 (2019), pp. 4–5.

50. IISD, "Summary of the 2023 Dubai Climate Change Conference: 30 November–13 December 2023," *Earth Negotiations Bulletin* 12, no. 842 (2023), p. 2.
51. IISD, "Summary of the 2023 Dubai Climate Change Conference: 30 November–13 December 2023," *Earth Negotiations Bulletin* 12, no. 842 (2023), p. 13.
52. IISD, "Summary of the 2023 Dubai Climate Change Conference: 30 November–13 December 2023," *Earth Negotiations Bulletin* 12, no. 842 (2023), p. 28.
53. UNEP, *Adaptation Gap Report 2023: Underfinanced and Underprepared. Inadequate Investment and Planning on Climate Adaptation Leaves World Exposed* (Nairobi: UNEP, 2023).
54. UNEP, *Adaptation Gap Report 2023: Underfinanced and Underprepared. Inadequate Investment and Planning on Climate Adaptation Leaves World Exposed* (Nairobi: UNEP, 2023), p. xv.
55. UNEP, *Adaptation Gap Report 2023: Underfinanced and Underprepared. Inadequate Investment and Planning on Climate Adaptation Leaves World Exposed* (Nairobi: UNEP, 2023), p. 30.
56. UNEP, *Adaptation Gap Report 2023: Underfinanced and Underprepared. Inadequate Investment and Planning on Climate Adaptation Leaves World Exposed* (Nairobi: UNEP, 2023), pp. 4–5.
57. UNEP, *Adaptation Gap Report 2023: Underfinanced and Underprepared. Inadequate Investment and Planning on Climate Adaptation Leaves World Exposed* (Nairobi: UNEP, 2023), p. 8.
58. UNEP, *Adaptation Gap Report 2023: Underfinanced and Underprepared. Inadequate Investment and Planning on Climate Adaptation Leaves World Exposed* (Nairobi: UNEP, 2023), p. 23.
59. UNEP, *Adaptation Gap Report 2023: Underfinanced and Underprepared. Inadequate Investment and Planning on Climate Adaptation Leaves World Exposed* (Nairobi: UNEP, 2023), p. 26.
60. UNEP, *Adaptation Gap Report 2023: Underfinanced and Underprepared. Inadequate Investment and Planning on Climate Adaptation Leaves World Exposed* (Nairobi: UNEP, 2023), p. 33.
61. I thank Paul G. Harris for this observation.

4

The Transformation Imperative

THE DEVELOPMENT OF ADAPTATION POLICY AND SCHOLARship has led to widespread embrace of transformation in some form or another as the necessary response to climate change.[1] "Growing recognition of the need for rapid and substantial change to address contemporary issues like climate change has led to the emergence of the concept of transformation," observe Ioan Fazey et al.[2] Transformation is required, various analysts argue, regarding worldviews, values, decisionmaking processes, economic sectors, institutions, and more.

What transformation means varies dramatically, from tweaking a local decisionmaking process to wholesale replacement of current global, national, and local institutions with new ones that support long-term sustainable development, equity, and democracy.[3] As temperatures rise, it is probable that the minor reforms will lose interest and the more extensive changes will take center stage. Indeed, that is already happening: the deputy secretary-general of the UN asserted recently, "This year will determine the future of the 2030 Agenda. We must achieve transformative change. World leaders will need to make a choice—to fulfill their commitment to a better future, or let it fall by the wayside."[4] Succinctly, prominent climate scientist Katharine Hayhoe states, "It's the system we all live in that must change."[5]

This chapter analyzes the meaning of transformation as a response to climate change. After outlining the growing consensus on the need for transformation, it discusses transformation in climate change scholarship. The next section assesses the view of the Intergovernmental Panel on Climate Change (IPCC) on transformation.

What's the Plan?

In 2014, David Spratt identified seven myths of the predominant climate policy-making paradigm.[6] They include:

- Climate change is not yet dangerous;
- The 2°C threshold is an appropriate policymaking focus;
- Tipping points are unlikely to initiate before temperature rises by 2°C;
- Mitigation should aim at 2°C, but planning should encompass adaptation to 4°C;
- A substantial carbon budget should remain below 2°C continues;
- Long-term feedbacks are not relevant to the carbon budget;
- Time remains to implement sufficient emission reductions while staying within existing political and economic institutions.

Since Spratt published this list of myths, the policy focus has changed to aim at 1.5°C, and the prediction of 4°C by 2100 has given way to a prediction of 2.8°C by 2100. The Paris agreement's pledges and some adjustments to the science have resulted in an expectation of a lower temperature rise than had been anticipated before Paris. Nonetheless, Spratt's analysis remains pertinent. For one, observers expect harms to occur at lower temperatures than once thought. Also, as Spratt observes, the CO_2 levels of 2014 were already higher than at any time since humans appeared on the planet. Today, the CO_2 and CO_2 equivalents level has reached 500 parts per million, 30 parts per million higher than when Spratt published. Consequently, big changes are certain to follow from current greenhouse gases (GHG) concentrations, no matter what policy path the world follows now. For example, when, in the distant past, GHG concentrations were at about the same level as those of today, the sea level was 20 to 30 meters higher. Further, at 2°C, well below the expected 2.8°C, it is likely that some tipping points will have been surpassed; the GHG concentration corresponding to 2°C has already been exceeded. Thus, the world is now witnessing melting of the permafrost, the rapid depletion of glacier ice, and the prospect soon of an ice-free summer Arctic Ocean. Ecosystems are transforming in response, with ecological ranges changing and extinction threatening a large share of plant and animal species. Many observers assert that human civilization cannot survive in a 4°C world, and perhaps not in a world of 2.8°C or higher; the limits of adaptation will have been reached. Thus, Spratt writes, "One question remains: if the world has practically speaking given up on holding to 2°C and it is not possible for human civilization to survive in a 4°C [or 2.8°C] warmer world, what's the plan?"[7]

One answer is to implement effective adaptation measures. But the record of adaptation policy reveals that current policy rests on unrealistic expectations that mitigation will succeed. Unless the most ambitious mitigation goals are met, incremental adaptation probably will not suffice. "Some projected impacts," notes the National Research Council, "are likely to be beyond the scope of adaptation, unless adaptation involves major structural change to government and society."[8] Thus, under any likely climate future, significant institutional change may be required to meet the challenge.

Climate scientists have asserted the current path toward 2.8°C or higher global average temperature by 2100 portends catastrophic effects on human civilization. Because nothing suggests the world will avoid that outcome, the question becomes whether global neoliberalism and the system of states afford a decisionmaking framework within which appropriate measures can be adopted and implemented to cope with a much warmer planet. If not, then current institutions must be counted as "hard limits" to successful adaptation to a much hotter climate, and the world must consider transformational change of global institutions to those that will enable effective, equitable adaptation to a hot world. Perhaps a different set of global economic, political, and social institutions can survive and maybe thrive in a world 2.8°C or more above the preindustrial global average temperature. If not, and if a hot world does mean the end of civilization as we now know it, then the world must prepare for a catastrophe of human making.

At present, not enough is being done, not even to respond to fairly low levels of global warming. A 2015 investigation of country reports on adaptation efforts found that only 23 percent of them "contained tangible adaptation actions."[9] Some countries reported no adaptation actions, remaining at the information-gathering stage, and one, Monaco, had no plans for evaluating its needs or enacting adaptation actions. Mainstreaming, which tends toward incremental efforts, was found to be the most common approach, whereas the climate crisis could become so acute that it requires transformative change, well beyond incorporating adaptation into plans for development and economic growth. Making progress on adaptation, said the study, would require better understanding and measurement of progress and more attention to adaptation efforts rather than preliminary evaluations and assessments. Similarly, Russ M. Wise et al. found that most adaptation projects are incremental, and the language of "adaptation pathways" limits vision for more effective alternatives.[10]

Today, participants and researchers find little consensus on important concepts or methods of measuring adaptation and transformation, and therefore lack the tools to assess whether climate policy is progressing at all, not to speak of initiating processes leading toward transformational change. In part, the lack of clarity arises from unwillingness to provide a picture of what kind of society would emerge from the process of transformational change. For the most part, the IPCC and other observers limit their comments to noting that transformational change means *fundamental* changes to systems, which only raises the question of what counts as fundamental. In the Sixth Assessment Report (AR6), the most recent IPCC report, adaptation is said to be part of climate-resilient development (CRD). This has the air of greater specificity, but if CRD is conducted within existing global social, economic, and political institutions, one must ask what about it is transformational. In short, furthering the discussion of adaptation requires close attention to the key concept of transformational change.

Consensus on the Need for Transformation

The term *transformation* appears in the Synthesis Report for the Fourth Assessment Report (AR4) only one time—in the glossary as part of the definition of structural change.[11] In 2014, the Synthesis Report for the Fifth Assessment Report (AR5) mentions transformation fourteen times, bringing it into the main text as an important option for coping with the effects of climate change, and transformation merits its own entry in the glossary. Further, chapters on regions, sectors, and responses highlight transformational adaptation and transformational change, reflecting the growing attention to transformation in official statements, the work of activist organizations, and scholarship. In the years since, calls for and investigations of transformation have become ubiquitous, emanating from the United Nations and participants in the policy process as well as activists and scholars. As often happens in policy and political discourse, the more a term is employed, the less it retains a determinate meaning, becoming one of those essentially contested concepts that embodies deep divisions over which course of action should be adopted. Nevertheless, the years since the publication of AR4 have seen a growing consensus on the need for some kind of transformation to meet the challenge of climate change.[12]

The UN Secretary-General asserted in 2021 that it was time to "embrace transformational change"[13] to meet a plethora of crises, including climate change, a call he has repeated many times since. Scholars working in the Institute of Development Studies published a study of transformative climate justice.[14] The ICF Climate Center's director asserted the need for "transforming systems and societies in order to move toward low-carbon, resilient development."[15] The Initiative for Climate Action Transparency has developed a measure of the transformative potential of mitigation policy.[16] The UN Development Programme has established an initiative for Scaling Up Climate Ambition on Land Use and Agriculture (SCALA) to contribute to transformative climate action, working in part with countries' National Adaptation Plans. The Mitigation Action Facility, originally founded by Germany and Great Britain and later joined by other donors, states that it supports transformational change through national mitigation plans.[17] A policy analysis published by the Prairie Climate Centre argues that cities should undertake transformational change to cope with the stresses of global warming.[18] The list of individuals, organizations, and studies researching, investigating, analyzing, and advocating transformation is all but endless. From public officials, private sector organizations, activists, and scholars, the call for transformation is on the rise. What is to be transformed and in what way? The answer is not altogether clear.

The Contested Meaning of Transformation

Transformation has been defined by various participants in the climate change policy dialogue in different ways.[19] The definitions share the notion that transformational adaptation is more extensive in scale and extent than incremental adaptation, but the line between it and incremental adaptation is blurry and context dependent.[20] Consequently, the definition of transformational adaptation is necessarily fluid, suggesting that its definition depends in part on the views of the observer. Transformational change is about systemic change in institutions, and it also lacks determinate content.

In 2013, a conference about transformation and climate change took place in Oslo. The Proceedings contain numerous articles noting the contested meaning of transformation to a low-carbon, well-adapted society. Understandings vary from transformation as instituting a green economy, one associated with "opportunities for innovation and increased levels of well-being," to the onset of "chaos and disruption," or the efforts of the powerful to benefit themselves from crisis situations. Clearly, "there are diverse values and interests at stake."[21] High stakes and divergent views do not lend themselves to consensus definitions. For instance, Karen O'Brien and Linda Sygna observe, "It is not always clear exactly what needs to be transformed and why, how, in whose interest, and what the consequences will be."[22] While most understandings highlight significant or fundamental change, in some cases transformation "can appear to have taken place without changing the actual underlying system."[23] Thomas M. Tanner and Aditya V. Bahadur note the "elasticity" in the term "reflected the shift in lexicon from adaptation to resilience and now to transformation," such that the term had been taken up in policy discussions "with little substance to guide implementation."[24]

In the years since the 2013 Oslo conference, the meaning of transformation has become, if anything, more open to interpretation. As the policy community has embraced the term, transformational adaptation and transformational change can appear supportive of current institutional arrangements. Indeed, transformation, by some accounts, requires the participation of private sector business enterprises, implying that corporations will continue to dominate the economic scene, pursuing profits and growth, albeit with an enhanced consciousness of the need for green practices. In official discourse, political transformation usually means enacting more inclusive processes to encourage community participation in adaptation planning, drawing on Indigenous knowledge and paying particular attention to women's voices. Apart from recommending and sometimes implementing the creation of new multilateral agencies to manage such processes, the structure of global governance would remain undisturbed.

Major environmental nongovernmental organizations (NGOs) have embraced the concept of transformation, but with varying understandings of its meaning. The International Union for Conservation of Nature (IUCN) has established the Nature 2030 Programme to encourage "transformative change" through five pathways: recognize, retain, restore, resource, and reconnect.[25] This initiative is aligned with the UN's Sustainable Development Goals (SDGs) and is intended to support conservation and respond to climate change. The World wildlife Fund (WWF) claims to be furthering the goals of the Paris agreement by encouraging "companies and cities to transform their businesses to dramatically reduce their emissions throughout their entire value chain."[26] That is, it counsels moving toward energy efficiency and nonfossil fuels without altering the structural features of the socioeconomic system. The WWF view is an instance of the widely embraced understanding of transformation as the transition to renewable energy. The World Resources Institute has published an extensive study identifying needed transformation across many sectors of the economy and society, and showing that the world is far from achieving sufficient transformation to head off ecological crisis, with transformation understood primarily as decarbonization.[27] In general, major NGOs pursue climate change action that supports the framework of the UN's Sustainable Development Goals.

Scholarship on Climate and Social Change

Some scholars have attempted to clarify and classify transformational responses to climate change. Typologies identify temporal, spatial, and systemic aspects of transformation, generally asserting that to be transformational there must be significant alteration of a given social-ecological domain in a fairly brief time span. Transformation can occur without intervention or planning, or it can be deliberate.[28] Generally, the purpose of identifying the factors driving transformation is to be able to control them so as to achieve a desirable outcome. Although scholars disagree on the means, most of them research and write on this subject because they seek a solution to the climate challenge, generally couched in terms of sustainability and equity.

Definitions of transformation in the climate research literature tend to employ such words as *fundamental, significant,* and the like to distinguish incremental adaptation and change from transformational. However, confusion can arise from conflating transformational adaptation with transformational change. Indeed, studying social transformation in response to ecological change, Giuseppe Feola notes that many uses of the term *transformation* have been metaphorical rather than analytical.[29] But correctly categorizing a case as transformational adaptation could make it more concrete and less metaphorical. For instance, Sarah F. Park et al. consider changes in the practices of Australian vintners to have been a transformational adaptation strategy, using various terms—*transition, transformational adaptation, transformative change*—more or less as synonyms.[30] But adopting forward-thinking business practices in one industry is not a case of transformational change, although it could be the transformational adaptation. Indeed, given the right context, the case might even illustrate fundamental or significant change in practices, albeit within a broader institutional structure that is largely untouched.

Another pitfall, seen in IPCC documents, is to assert the need for transformation while recommending incremental reforms of existing practices. For example, Frank Biermann et al. usefully focus on institutions, not adaptation projects and measures, but their call for transformation of institutions leads only to several suggestions for reforms of existing international organizations. The authors declare incremental change to be insufficient, and assert the need for "fundamental reorientation and restructuring of international institutions toward more effective Earth system governance and planetary stewardship."[31] Yet their suggestions are that international actors should raise the prominence of sustainable development in the UN system, fill regulatory gaps, mainstream environmental goals into economic regimes, ensure greater transparency, simplify decisionmaking procedures, and provide ample financial resources to developing countries. These are all useful suggestions, but they fall short of constituting "transformation," which must mean more than reforms that assume the continuation of current institutions of global governance. Similarly, one study states that the shift to "adaptive governance" is transformational, but it is expected to occur within a preexisting legal structure.[32]

Much of the scholarship on transformation provides little clear idea of what kind of society would result from transformational change. Carl Folke et al. define the term *transformability* as the capacity to move to "a fundamentally new system when ecological, economic, or social structures make the existing system untenable."[33] Karen O'Brien writes, "Transformation can be defined as physical and/or qualitative changes in form, structure or meaning making," as well as a cognitive change toward care, and it can imply both opportunity and risk. Transformation can result from conscious effort to bring about change, and from unintended or unexpected processes and events. O'Brien calls for a "science of deliberate transformation" that would entail "a questioning of the assumptions, beliefs, values, commitments, loyalties and interests that have created the structures, systems and behaviours that contribute to anthropogenic climate change, social vulnerability and other environmental problems in the first place."[34] Robert W. Kates, William R. Travis, and Thomas J. Wilbanks suggest transformational adaptations fall into three classes: those of large scale and intensity, those truly new to a region or system, and those to places and shifts of location. In each category, transformations encompass technological, behavioral, and institutional changes.[35] Samuel Wilson et al. understand socioecological system (SES) transformation as "the process of deep change of identity (goal), feedback processes, structure, and functions."[36] Michele-Lee Moore et al. distinguish transformation from adaptation, asserting that the former "recombines existing elements of a system in fundamentally novel ways."[37] Clive A. McAlpine et al. insist that global society must move "the importance of intergenerational equity and environmental sustainability to the top of the political agenda and to the core of personal and societal belief systems," which, in turn, will entail "major adjustments to social and economic systems."[38] Ines S. Campos et al. argue that transformation refers to the transition to sustainability that can be investigated in terms of participatory action-research; case studies in Portugal show how to promote sustainable adaptation processes and guide research into adaptation.[39] In general, these works focus on cognitive and value changes without specifying the institutions and social practices that are to follow.

Mark Pelling's influential work on transformation and climate forwards a primarily cognitive view of transformation. Pelling distinguishes resilience, transition, and transformation in the process of adaptation. Transformational adaptation relates to "notions of risk society, the social contract and human security,"[40] although other theoretical perspectives can also be brought to bear. For Pelling, "transformational adaptation describes those actions that can result in overturning of established rights systems and the imposition of new regimes."[41] Moreover, "in identifying the assumptions that underlie modernity as a potential focus for adaptation[,] transformation is also directed towards internal-cognitive change." Pelling argues that incorporating the views and interests of future generations and people in other countries into adaptation policy can result in "transformational adaptation" that is what leads to the overturning of established rights systems and the imposition of new regimes. Further, transformational adaptation can be understood in terms of which system is involved, the observer's viewpoint, and intention, action, and outcome. In sum, transformation begins with cognitive change and results in replacement of existing rights systems with unspecified new ones.[42]

Similarly, Petra Tschakert and Asuncion Lera St. Clair call for "a radical notion of transformative change" rooted in "a fundamentally different conceptualization of the world."[43] The main elements of this conceptualization are a relational ontology of responsibility and care, embedded enactments of care, and an understanding of vulnerability that emphasizes interdependence. The "radical" aspect of this framework lies in a different way of understanding the world and how people, even at a distance, can relate morally. This argument, drawing in part on postcolonial theory, offers some useful observations. For one, its reinterpretation of vulnerability shifts the frame from the supposedly self-reliant developed nations providing charitable aid to the weak, vulnerable poor and developing countries. Instead, vulnerability is more aptly framed as an existential reality for everyone, constituted in interactions and relationships with others. This helpfully challenges the tendency to relegate such problems as adaptation to climate change to distant places affecting unfortunate people.

Yet Tschakert and St. Clair do not offer much by way of practical guidance on effecting radical transformation. Their analysis remains at an abstract level, and the radical change is mostly to occur in consciousness, with hoped-for material results that are not all that radical. Examples of supposedly radical transformation include ethical consumption, transition towns, and permaculture, all of which fit quite comfortably into global capitalism, local action, and individual behavior. In the one significant practical change suggested, the authors close with a call for "intentional and deliberative transformation toward equitable, low-carbon societies."[44] Thus, transformation here is about the energy grid, and, as with other analysts, amounts to switching to renewables. This, of course, leaves all the difficult political questions unmentioned, much less analyzed. It ignores that suggested pathways to low-carbon society range from those entirely reliant on market mechanisms and the profit motive to calls for ecosocialism. The reader is left to infer which outcome follows from a relational framing of climate change, recognizing human and nonhuman moral and practical interdependence.

The point of discussing this particular article by Tschakert and St. Clair is not to denigrate the important contribution the authors make. It is to highlight the problems with cognitive orientations to transformation, an orientation that is found across the scholarship on climate change adaptation. Focusing on ideas, values, and worldviews is important, of course, but doing so has limits. For one, changing how individuals feel about the world can be impotent if not connected to a project for transforming social and economic institutions. Further, as Alexander Wendt notes, changing deeply embedded social constructions of reality can be far more difficult than changing material circumstances, so it is not entirely obvious that transforming worldviews is the place to start. The cognitive structures of domestic and international society incline actors to reproduce them, and social facts "might not be malleable in some historical circumstances."[45] In today's world, deeply embedded social constructions that stand in the way of transformation toward ecological sustainability include militarized sovereign states, capitalist property relations, and global consumerism.[46] It may well be that the prerequisite for bringing about transformative cognitive change toward an ethic of care on a global scale requires first upending the institutional arrangements that stand in the way of that transformation. If so, then prevailing ideas follow material change rather than preceding it.

Calls for cognitive transformation that do not confront structures of power and privilege and that do not recognize how resistant they are to change are not serious. No doubt, behavior depends on belief, and so enlightened cognitive understandings of adaptation are valuable. Yet scholarship on cognitive transformation must recognize that an ethic of care or other green orientations will have little efficacy against long-standing and entrenched worldviews supporting military competition, capital accumulation, and endless growth. There is little reason to expect education and exhortation to transform international politics soon enough to prevent global warming beyond the internationally agreed 2°C threshold for dangerous impacts. More to the point, embracing ecological values, beliefs, and knowledge only sets the stage for confronting the institutional arrangements that thwart the translation of those cognitive orientations into practical outcomes. There is no avoiding the central question of what must be transformed, into what, and how.

More concrete proposals suggest that transformation might occur by way of attaining the Sustainable Development Goals. For that to occur, assert Albert V. Norstrom et al., the process of setting SDGs should encompass an integrated socioecological perspective, recognition of trade-offs between ambition and feasibility, and knowledge "about the principles, dynamics, and constraints of social change processes at all scales, from the individual to the global."[47] This view is in line with the IPCC's most recent analysis of the topic, discussed below.

Definitions of transformation differ and talk past one another because scholars bring significantly different views regarding the potential for existing institutional arrangements to bring about sufficient change in policies and practices to meet the climate challenge. Making sense of this growing body of literature presents daunting challenges. But beyond parsing definitions, comprehending the meaning of transformation requires an idea of the social, economic, political, and cultural institutions that are expected to replace current arrangements. Without some indication of where the scholar, activist, or policymaker stands on that question, one cannot know whether they are indeed calling for transformative change or simply business as usual with a green hue.

This is not to say that the concept of transformative change points in only one direction. To the contrary, many transformative changes in global institutions are possible and will be reviewed in the next chapter. But not just any change counts as transformative. The term is contested, to be sure, but it is not meaningless.

Classifying Transformations

Scholars have offered several attempts to clarify the concept of transformation in the context of climate change. The most ambitious attempt might be Guiseppe Feola's mapping of the conceptual and methodological diversity of the transformation literature.[48] Feola conducted a wide-ranging survey of literature on ecological transformation and classified works along four dimensions: system model, form and temporal range, seat of causality and social consciousness, and outcome. This conceptual map yields seven orientations to transformation:

- Progressive transformation: reform or replacement of existing social contracts to tackle vulnerability at its roots;
- Regime shift: initiating change at lower scales to bring about system change without sacrificing system resilience;
- Societal transition: multisector and multiactor changes in long-term processes at the macro level;
- Social practice: reconfiguration of existing practices;
- Transformational adaptation 1: large scale, intensity, novelty, and spatial change;
- Transformational adaptation 2: change in systems that enhances the capacity for desired values to be achieved;
- Socioecological transition: changing the throughput of energy and material in the system.

These perspectives share the assumption of a unified or interacting system encompassing human, technological, biological, and natural elements. They differ in regard to how the components relate, including which are fundamental to understanding the system. All call for structural change, although what that means can vary widely. Some are concerned with the individual level while others stay at a macro or systemic level of analysis. They also differ as to whether local changes can properly be deemed transformational. All these perspectives, according to Feola, view transformational change as "characterised by discontinuities, ruptures, or thresholds, and do not generally proceed smoothly."[49]

In addition, the perspectives vary on whether they see the possibility of directed or managed change. Those with historical views of transformation see it as "emergent," the product of long-term social changes such as the transition from feudalism to capitalism. Others would deem the transition from fossil fuels to renewable energy as transformational, a process that can be guided by policy and accomplished in decades, not centuries. Some orientations to transformation tie it to movement toward a desired outcome, such as fulfillment of the Sustainable Development Goals, while others are agnostic about the outcome.

Feola's mapping of literature on transformation and ecology usefully clarifies how scholars understand the concept. It does not attempt to resolve differences to provide a coherent definition of the term *transformation*. Nor does Feola note one of the important absences in this literature: the lack of elaborated visions of what comes next. We are asked to think differently, but few scholars suggest where changed values and worldviews should lead. More often, studies that discuss transformation assert the need for more research on how societies transform.

Neglect of Large Bodies of Research on Social Change

Studies of transformation and the climate often include calls for more research into the drivers of social transformation. However, few acknowledge large bodies of literature on just this question, including classic works and peer-reviewed publications, as well as extensive contemporary writings on social change. Reference lists are remarkable for the absence of such names as Karl Marx, Max Weber, and Emile Durkheim.[50] One finds little to no mention of Immanuel Wallerstein, John Bellamy Foster, Leslie Sklair, William I. Robinson, or Giovanni Arrighi, to mention a few. This is not to say such viewpoints are entirely unrepresented in the climate literature, but no study of transformation should call for more research on the topic while ignoring nearly two centuries of work on the process of social change.

Taking account of critical voices in the broad stream of social theory and research would point to another major absence in much of the scholarly work on climate and social change.[51] That is, there are innumerable calls for moving toward sustainability, equity, and inclusiveness. However, few scholars indicate what institutional and practical changes would be required to bring about the just and sustainable world to which they aspire. As noted, many studies leave the question at fostering ecological consciousness and the assumption that broad changes in beliefs and attitudes will lead to unspecified institutional transformations. The only frequently mentioned change of practices is to transition from fossil fuels to renewables, along with perhaps improved land use, to reduce GHG emissions.

Yet significant and widely known work on social change exists. For example, Karl Marx, Friedrich Engels, and numerous social thinkers following them, have explained that class relations are the fundamental element of social change, the "base" driving changes in the "superstructure" of ideas, politics, and culture. The profit motive, exploitation of workers, and pillaging of natural resources are prominent characteristics of capitalism, according to these scholars. Social change is the result of the developing contradictions of a capitalist society, driven in part by technological change, in part by the heightening of class conflict. Transformation occurs when the contradictions can no longer be contained within the institutional framework of capitalist class relations, and broad opposition movements replace the fetters of capitalism with a novel social order. Consequently, transformation means putting society on an entirely new basis; namely, a foundation of socialism. In today's dialogue, scholars and activists in this tradition call for ecosocialism or ecological civilization. This orientation to social change has been present in social theory for more than two centuries, and the paradigm is alive today, represented in numerous books, journals, and other publications, as well as conferences and organizations. The volume of writings and research on social change and transformation is immense, and recent work devotes considerable attention to ecological challenges. Yet mainstream research on climate change, including that informing official reports, suggests the need for more research on social change without acknowledging the vast literature on the topic that is already available.

These absences in scholarly literature are reflected in the IPCC's assessment reports, which are, significantly, based on peer-reviewed literature. The IPCC has raised the level of normative ambition in its shift from a technocratic and managerial understanding of adaptation to an embrace of climate-resilient development. Yet it has done so while avoiding questions of institutional change.

The IPCC and Transformational Change

The IPCC offers the most influential effort to bring scientific and scholarly research to bear on climate change policy. The IPCC has forefronted transformation in response to global climate change, but it notes that even within AR6, the term *transformation* can be employed in various ways. The organization states, "As used in the global environmental change literature, transformation is a pluralistic concept embracing many interpretations (Box 18.3), but all focus on the general idea of fundamental change in society as opposed to change that is minor, marginal or incremental."[52] However, what counts as "fundamental" remains contested.

The review of adaptation in climate negotiations presented in the first three chapters of this book indicates that the IPCC has increasingly emphasized transformation to respond to the climate challenge. In AR5, transformational adaptation and transformational change were posited as required to address the impact of global warming on nature and human interests.[53] In AR6, the IPCC makes a clear and unequivocal shift toward transformation as a moral and ethical necessity to achieve justice in a sustainable world. Yet the same absence of specifics about institutions found in scholarly literature appears in AR5 and AR6. Consequently, the IPCC's potentially radical prescriptions give way to an implicit expectation that extant institutions, the system of global governance and global capitalism, can and must carry out the needed transformations in various sectors of the economy and society.

Increasing Normative Ambition

The increasing normative ambition of the IPCC appears in a comparison of the Assessment Reports from the first in 1990 to the most recent concluded in 2023, showing considerable shifts in orientation. In the scientific community's first cuts at the implications of climate change, it was more or less assumed that GHG emissions would continue, temperatures would rise, and the main question was how and how well human societies would adjust and adapt to a significantly different global climate, understood largely as a technical and management problem. In recent reports, adaptation is seen as an essential component of achieving sustainable, equitable development, working hand in hand with mitigation to prevent the impacts of climate change from disrupting civilization as we know it.

The IPCC, in AR6, asserts that climate-resilient development requires justice, inclusiveness, and compatibility with human rights. The AR6 Summary for Policymakers begins with a statement of the report's orientation: "This report recognizes the interdependence of climate, ecosystems and biodiversity, and human societies; the value of diverse forms of knowledge; and the close linkages between climate change adaptation, mitigation, ecosystem health, human well-being and *sustainable development*, and reflects the increasing diversity of actors involved in climate action."[54]

As one might expect from a UN body, the IPCC intends for its policy recommendations to contribute to fulfillment of the Sustainable Development Goals. The SDGs represent a higher level of ambition for global social, economic, and political change than the predecessor Millenium Development Goals, and recommendations for climate policy conform with that tendency.

AR6 forwards the notion of climate-resilient development as the path toward an effective response to the climate challenge. The IPCC asserts, "Prioritising equity, climate justice, social justice, inclusion and just transition processes can enable *adaptation* and ambitious mitigation actions and climate resilient development."[55] The IPCC, in line with the scholarly literature on which its reports are based, finds that the greatest adaptation challenges lie in the developing world, and the solution is to fulfill the SDGs while recognizing that climate change can be a barrier to doing so. The suggested remedies include enhanced international cooperation and more money for adaptation, with no hint of significant institutional change.

In what sense is the IPCC exhibiting greater normative ambition? For one, the IPCC has fully embraced the value of equity as desirable in its own right and essential to formulating and implementing effective adaptation. In addition, the IPCC expresses support for more inclusive, open decisionmaking processes, albeit with ample input from scientists and experts to inform public debate. The IPCC also increasingly emphasizes the need to listen to Indigenous people on human-nature relations, and it insists on respect for women's rights, again claiming that gender equity is essential to effective adaptation policy.

Viewed broadly, the IPCC's reports have addressed several basic questions. First, is global warming happening? Second, if so, what is causing it? Third, what effects will it have on natural and social systems? Fourth, what can be done? It then surveys reputable sources (defined mainly by appearance in peer-reviewed scholarship) for answers to these questions. In that regard, the IPCC can lay claim to a degree of detachment from the subject, maintaining scientific objectivity. However, it has become evident that important concepts that observers employ to analyze adaptation to climate change are products of social and political processes that are normatively charged. For instance, determining what counts as vulnerability depends on prior assumptions about what and who is threatened, and whether it matters.

The IPCC has resolved deeply political and normative questions by adopting sustainable development as its guide, presumably answering the question of what is to be done in terms accepted and promoted by most of the world's governments and major international organizations. Consequently, it can then assert that a successful CRD response to climate change requires enhanced international cooperation, greater funding for adaptation in developing countries, and democratic decision-making processes; in short, a stepped-up version of the traditional development agenda. Its claim to objectivity stems from adopting the objectives of others; namely, the United Nations as expressed in the SDGs, rather than forwarding its own preferences.

Nevertheless, these are political and normative choices, not necessary conclusions drawn from climate science. To be sure, the heightened sense that justice and democracy are requisites for effective climate policy arises from the perception that the effects of climate change are falling on the weak and poor, those effects are becoming increasingly evident, and the powerful and wealthy have shown little inclination to halt the growth of GHG emissions. But climate science does not provide a definitive answer to how to respond to that situation, even assuming all actors have accepted that it is basically correct. The IPCC's analysis reflects a consensus among the authors that certain values in line with the SDGs must be incorporated in climate policy; these values are part of the definition of successful adaptation, and necessary conditions for achieving it.

Yet although the IPCC takes a stand on how to remedy the effects of climate change—that is, how to adapt successfully—once again, it pulls up short of the potentially radical implications of its recommendations. A full chapter of the report of Working Group II is devoted to explaining and recommending CRD, defined as the successful integration of mitigation, adaptation, and sustainable development.[56] The chapter asserts that current development pathways are leading away from CRD, made worse by such trends as rising income inequality, the increase in greenhouse gas emissions, unhelpful land use changes, food and water insecurity, forced migration, and reversals regarding human life expectancy. These tendencies, state the authors, worsen poverty, injustice, inequity, and environmental degradation, which climate change can exacerbate.[57] To correct this and get on the path to CRD will require significant change. System transitions in "energy, industry, urban and infrastructure, land and ecosystems, and societal" are required to enable CRD and help avoid crossing planetary boundaries, if the appropriate enabling conditions and inclusive engagement are also present. These system transitions can, in turn, lead to transformations, understood as "deeper and more fundamental climate changes than transitions, including new underlying values, worldviews, ideologies, structures and power relationships.[58] But focusing on the last two, what are the structures and power relationships in need of fundamental change, and what are they to change to? The IPCC is not forthcoming on these questions, perhaps because an honest appraisal would constitute a critique of wealthy and powerful interests represented in the governments that oversee the IPCC's work.

Observers, analysts, and participants in climate policy cannot help but take a stand. Consequently, the IPCC's recent reports show growing emphasis on potentially controversial political values rather than the technical approach it had taken in earlier discussions. The IPCC argues that the requisites for CRD include giving priority to "equity, climate justice, social justice, inclusion and just transition processes."[59] Its formulation and advocacy of CRD could give direction to climate policy, perhaps encouraging the coherence of policy emerging from many levels, regions, and sectors that it now lacks. But taking that stand also invites criticism of the body as politicized and normatively wrong.

Institutions Compatible with Transformational Change

Roger A. Pielke Jr. and others critique the IPCC for hiding a left-wing political project behind a purportedly objective presentation of ecologically sound and desirable policies. According to this view, "transformational change" has become the central symbol of a broad effort to enact progressive, perhaps socialist, values. However desirable changing everything might be, advocating it, argues Pielke, is not the appropriate role of a body charged with providing neutral scientific information to policymakers and the public. Rather, it represents the capture of the IPCC by an in-group with shared political views; namely, a commitment to degrowth, which will "turn climate policy on its head such that ends become means." Consequently, the "IPCC has clearly departed from its role as a scientific assessment and is now much more deeply engaged in political advocacy."[60]

This critique misreads the IPCC and its major reports: for the IPCC assumes that climate policy will be formulated and implemented through existing governments at all levels, and both civil society and the private sector will play crucial roles in the transition to CRD. While acknowledging that ambitious mitigation pathways can bring "disruptive changes in existing economic structures," the IPCC's recommendations are tepid: appropriate macroeconomic policies and strong social safety nets, hardly the stuff of socialist revolution.[61] In thousands of pages of IPCC reports, written by a large number of scholars and scientists working in a wide array of academic fields, one can no doubt find passages suggesting the desirability of degrowth and the like, but the thrust of the IPCC's work is to support development, including alleviating poverty in the developing world through robust, if green, economic growth.

Indeed, in a section on defining transformation, AR6 deems a plan developed by the Global Commission on the Economy and Climate as transformational,[62] whereas its recommendations include such standard items as higher energy efficiency, compact urban areas, reforestation, and investments in infrastructure, all aimed at "growth that is strong, sustainable, balanced, and inclusive."[63] The IPCC also considers Kate Raworth's doughnut hole economics, built on the planetary boundaries concept, to be transformational. Raworth locates the shortcomings of current approaches in poor policy choices, bad advice from mainstream economics, and insufficient implementation of international commitments. To ensure that human society remains below planetary limits and above human deprivation limits would require such things as respect for reproductive rights, better insulation of homes, reductions in food waste, and higher energy efficiency. The focus is on major policy shifts, presumably to be taken by governments, corporations, and other actors populating the current global scene.[64] Again, the transformational nature of these recommendations is hard to discern.

Nor does the IPCC simply advocate degrowth. Summarizing the literature, the IPCC asserts that there are two important perspectives on how to achieve the SDGs and fulfill the Paris agreement: ecomodernism and degrowth. Ecomodernism seeks to employ market incentives to encourage energy efficiency and other steps toward environmental protection. Degrowth, says the IPCC, would rely on redistribution instead of technological progress and growth in gross domestic product (GDP). Contrary to Pielke, the IPCC does not declare which is preferred, simply noting that the dialogue between the two camps has generated "mutual counterarguments."[65] Thus, even when the potential for choices about institutions is presented, the IPCC shrinks from taking sides, and it certainly does not covertly push a left-wing agenda.

The IPCC has not adopted radical social theory as its guide to the future, at variance with the perspectives and priorities of nearly every Member State of the United Nations. Rather, the IPCC ties its analysis and more or less implicit recommendations to the Sustainable Development Goals, which, it is important to note, have been adopted by nearly every UN member as the aspirational purposes of the organization looking ahead to 2030. And when the SDGs expire, it is highly likely that another set of internationally negotiated goals will be adopted, just as the SDGs succeeded the prior Millennium Development Goals. If the IPCC still exists, it will likely align itself with that international normative consensus, hammered out in negotiations among the hard-headed diplomats and other official representatives of national governments, not by "watermelon" (green on the outside, red on the inside) academics and activists. The IPCC is not, then, a rogue entity, unmoored from established political institutions and pressing for a radical agenda favored only by a small number of leftist advocates. To the contrary, for all its reiteration of the need for transformational change, the IPCC displays a distinct lack of vision as to what that might mean. The content of transformational change is left to the imagination, or to the political arena where states, corporations, NGOs, and social movements can fight it out, ensuring that structures of privilege, wealth, and power remain largely undisturbed.

To the point, the IPCC states, "CRD is not a predefined goal to be achieved at a certain point or stage in the future. It is a constant process of evaluating, valuing, acting and adjusting various options for mitigations, adaptation and sustainable development, shaped by societal values as well as contestations of those values."[66] The IPCC hopes that respecting variations in the policy context will ensure that the priorities will be "people, planet, prosperity, peace and partnership." But leaving the institutional outcome to future political struggles provides scant foundation for this hope. In particular, the chapter on CRD contains no mentions of *capitalism* in the text, and only four works using the term appear in over fifty pages of references. More than a few observers argue that global capitalism is a significant barrier to prioritizing CRD and the SDGs. Accordingly, a study aimed partly at identifying and suggesting remedies for barriers to adaptation and to CRD avoids pointing at the proverbial elephant in the room.[67] Certainly, a fair evaluation might conclude that, on balance, achieving CRD within the context of global capitalism is feasible and optimal, but the argument should be made, not quietly assumed.

Reformed and Augmented Liberal Political and Economic Institutions

Lack of clear vision regarding future institutional arrangements suggests the IPCC is far from radical and to the contrary advocates reforms to existing social, economic, and political arrangements. Governments are expected to provide adequate social safety nets, along with appropriate policies, regulations, and incentives. A properly reoriented private sector would adapt its operations to cope with climate change, provide insurance against climate risks, and contribute to financing adaptation measures, while profit-seeking private enterprise would carry on. Communities and individuals would play important roles in designing and implementing such responses as ecosystem-based adaptation, guided by experts and professionals. International organizations such as the United Nations would help coordinate adaptation actions and foster values supporting CRD and the SDGs. Transformation, it appears, is about infusing the existing structure of global governance at all levels with cognitive orientations that would lead to greater support for CRD. In practical terms, as we have seen, this means primarily transitioning to nonfossil fuel energy and altered land use practices. The institutional structure itself would remain largely untouched.

Indeed, a chapter in AR6 Working Group II reverts to the technocratic language of "climate risk management" to analyze the requisites for making decisions about climate policy.[68] The authors report that the available adaptation and climate risk management options include nature restoration, changing diets and reducing food waste, infrastructure retrofitting, appropriate building codes, disaster early warning, and cooperative governance. The radical content of this set of options is slight. Presumably, cognitive change will ensure that sufficient support exists to lead the various climate risk management options toward the best decisions and policies available within the given global order. This would not be a heavy lift, as it avoids challenging the broad social context of a civilization built on a growth-oriented, profit-seeking economic system and all the institutional arrangements—from family structure to the educational system to understandings of security—associated with it.

The most promising transformational options, says the IPCC, are migration, spatial planning, governance cooperation, universal health care, and modified food systems.[69] Of these, the one that is clearly transformational (at the community and national level) is migration, but as discussed below, it might properly be deemed a form of severe maladaptation rather than the kind of transformational adaptation most observers would prefer. Meanwhile, the IPCC asserts, "National and international legal and policy frameworks and instruments support the planning and implementation of adaptation and climate risk management across scales, especially when combined with guidelines for action."[70] Transformation will have occurred, then, when national and international authorities adopt and implement appropriate guidelines regarding migration, zoning, polycentric cooperation, access to health care, and sustainable agriculture. Of course, the requisite financing must be made available from public and private sources, both of which are currently inadequate. In short, transformational change is proper guidelines adopted by existing authorities and backed by sufficient funding.

Some hints at a more ambitious agenda can be found in IPCC reports. Yet they often operate at a high level of abstraction, leaving out the content needed to know whether the IPCC envisions significant institutional change. For example, AR6's chapter on decisionmaking identifies "three broad enabling conditions" for adaptation planning and implementation: governance, finance, and knowledge.[71] No doubt, for these three enabling conditions could define any decisionmaking context and could lead anywhere. The reader might provide the understanding of the current world order and the changes that must be made to achieve the goals that the IPCC suggests, but the IPCC does not do so itself.

Instead, the IPCC (and most of the academic literature on which it relies) sees the continuation of the current structure of world affairs, albeit infused with new values that support the transition to renewable energy, better land use practices, and a variety of other adaptation responses. This shows up in Table 17.1 of AR6 on Working Group II, in which eight broad areas of risk and three adaptation options for each area are identified, with a number of examples given for each adaptation option.[72] The table moves from the abstract to the specific. As a study of observed adaptation decisionmaking, it is not surprising that the options and examples show what current institutions can do. Nevertheless, the list of adaptation options reveals the limited vision contained in the IPCC's analysis. For example, in the area of risks to living standards and equity, the measures include insurance, diversification of livelihoods, and stronger social safety nets. These options might be desirable and optimal in current circumstances, but they are not transformational. Adoption and implementation of such responses to climate change is well within the capacities of existing institutions, although the record of progress on such fronts is spotty. Truly transformational changes to address threats to livelihoods, such as universal adoption of worker-owned enterprises, are not considered. In such ways, the IPCC, reflecting scholarship and the views of states and other significant international actors, eschews opportunities to describe and perhaps recommend transformational change that would enable the global community to choose a sustainable, resilient, equitable climate future.

Thwarted Transformational Potential

The IPCC states in many places that the barriers to effective adaptation to climate change include inequality, discrimination, economic practices, and power structures. For example, the IPCC asserts that vulnerability to climate change results from "patterns of intersecting socioeconomic development, unsustainable ocean and land use, inequity, marginalization, historical and ongoing patterns of inequity such as colonialism, and governance."[73] This inventory of barriers might lead to recommendations of transformational change in the nature of global economic relations such as building a new system that does not rely on the profit motive, does not require endless economic growth, and democratizes workplace ownership and decisionmaking. Instead, the IPCC avoids describing and explaining why current economic and power arrangements stand in the way of climate policy. It fails to elaborate which economic and power arrangements would enable, rather than thwart, CRD, or how to bring about transformation of those arrangements. Instead, the IPCC reports that adaptation actions are under way worldwide, although these efforts are limited and incremental, encompassing such initiatives as "decision support tools and climate services" and "pilot projects and local experiments."[74]

More must be done and the IPCC indicates what that, in its view, entails. It is not a transformation. Here is one the clearest statements of the IPCC's adaptation agenda:

> Integrating climate adaptation into social protection programs, including cash transfers and public works programmes, is highly feasible and increases resilience to climate change, especially when supported by basic services and infrastructure. Social safety nets are increasingly being reconfigured to build adaptive capacities of the most vulnerable in rural and also urban communities. Social safety nets that support climate change adaptation have strong co-benefits with development goals such as education, poverty alleviation, gender inclusion and food security.[75]

That is, government programs with adequate funding (including development assistance) are important parts of the climate policy toolbox.

What stands in the way of this limited set of recommendations? Why are they not already incorporated in national and international climate policy? The IPCC responds that the "soft limits" to effective adaptation are "primarily financial, governance, institutional and policy constraints."[76] The constraints include poverty, inequity, lack of climate literacy, inadequate information, and the inadequacy of financial flows for adaptation. Presumably, governments working with international organizations, the private sector, and NGOs can correct these shortcomings.

But what are the sources of poverty, inequity, illiteracy, and inadequate financial flows? What are the social, economic, and political structures that produce the constraints on policy and action? Do IPCC recommendations address the roots of the problem? It is not hard to see that the system of sovereign states and the global neoliberal economy are the barriers, the very arrangements that give status to the world's most privileged and powerful classes. The "realist" model of international affairs that focuses attention on the power and interests of states as the drivers of international action is the stuff of college introduction to international relations courses, yet the IPCC ignores this basic and accessible understanding of world affairs, preferring an obscure language of linked socioeconomic systems and risk management theory. Likewise, knowledge of the global political economy is readily available, if controversial. Yet the Summary for Policymakers of Working Group II does not employ the word *capital,* not to speak of *capitalism*. Thus, the financial and policy "constraints" standing in the way of effective adaptation go unidentified and are not analyzed. Meanwhile, one finds numerous calls for more research into the process of social change with little to no recognition of vast bodies of literature on that very topic, in classic and in contemporary studies.

This is not to say that the IPCC's analysis and the literature on which it is based do not point in the right direction. After all, the IPCC is intergovernmental, so it is no surprise that the authors, checked by government representatives, pull up short of spelling out the radical implications of their findings on the causes, effects, and responses to climate change. It is no surprise that the IPCC would limit itself to suggesting current structures of global governance will enact the needed policies, and that cognitive reorientations are to be the motor of changing policies. But scholars and citizens need not limit their vision to what states consider acceptable, states that too often represent the interests of their most powerful groups rather than the popular will. Scholars and citizens can draw the conclusions regarding global political and economic power and privilege left unstated in IPCC reports. Taking the next step from abstraction to a program of action requires fulfilling the thwarted transformational potential of the IPCC's analysis.

In sum, attention to transformation has been increasing in scholarship, environmental activism, and policymaking circles. Yet a clear understanding of what transformation means is lacking. Many definitions have been forwarded, sharing a broad notion of "fundamental" changes required to formulate and implement successful adaptation to climate change. However, most analysts and advocates stop short of specifying the institutional changes that would follow from cognitive change and broad acceptance of values supporting transformation to an ecologically sustainable society that can equitably and effectively adapt to climate change, nor, for that matter, a society able to achieve CRD or fulfillment of the SDGs. Without elaboration of the outcome, there is no certain way to know whether the recommended measures would lead to, or themselves constitute, transformation. Calls for more research on processes of social change mostly neglect large bodies of literature on that very topic. But numerous studies that do take the necessary next step exist, illuminating the challenge of transformation change. They are the main topic of Chapter 5.

Notes

1. Lukas Hermwille, "Climate Change as a Transformation Challenge: A New Climate Policy Paradigm?" *GAIA—Ecological Perspectives for Science and Society* 25, no. 1 (2016): 19–22; Mark Pelling, Karen O'Brien, and David Matyas, "Adaptation and Transformation," *Climatic Change* 133, no. 1 (2015): 113–127; Catrien J. A. M. Termeer, Art Dewulf, and G. Robbert Biesbroek, "Transformational Change: Governance Interventions for Climate Change Adaptation from a Continuous Change Perspective," *Journal of Environmental Planning and Management* 60, no. 4 (2017): 558–576.

2. Ioan Fazey, Peter Moug, Simon Allen, Kate Beckmann, David Blackwood, Mike Bonaventura, and Kathryn Burnett et al., "Transformation in a Changing Climate: A Research Agenda," *Climate and Development* 10, no. 3 (2018): 197–217.

3. Katharina Hölscher, Julia M. Wittmayer, and Derek Loorbach, "Transition Versus Transformation: What's the Difference," *Environmental Innovation and Societal Transitions* 27 (2018): 1–3.

4. UN Deputy Secretary-General Amina Mohammed, "Transformative Change Needed to Get the Global Goals Back on Track," *UN News*, April 27, 2023.

5. Katharine Hayhoe, *Saving Us: A Climate Scientist's Call for Hope and Healing in a Divided World* (New York: One Signal, 2021), p. 150.

6 David Spratt, *Dangerous Climate Change: Myths and Reality* (Melbourne: Climate Code Red, 2014).

7. David Spratt, *Dangerous Climate Change: Myths and Reality* (Melbourne: Climate Code Red, 2014), p. 6.

8. National Research Council, *Adapting to the Impacts of Climate Change* (Washington, DC: National Academies Press, 2010), p. 4.

9. Alexandra C. Lesnikowski, James D. Ford, Lea Berrang-Ford, Magda Barrera, and Jody Heymann, "How Are We Adapting to Climate Change? A Global Assessment," *Mitigation and Adaptation Strategies for Global Change* 20, no. 2 (2015): 277–293, at 283.

10. Russ M. Wise, Ioan Fazey, Mark Stafford Smith, Sarah E. Park, Hallie C. Eakin, Emma R. M. Archer Van Garderen, and Bruce M. Campbell, "Reconceptualising Adaptation to Climate Change as Part of Pathways of Change and Response," *Global Environmental Change* 28 (2014): 325–336.

11. Intergovernmental Panel on Climate Change (IPCC), *Climate Change 2007: Synthesis Report. Contribution of Working Groups I, II and III to the Fourth Assessment Report of the Intergovernmental Panel on Climate Change* (Geneva: IPCC, 2007), p. 87.

12. Laurie Laybourn, Henry Throw, and Suzannah Sherman, *1.5°C—Dead or Alive: The Risks to Transformational Change from Reaching and Breaching the Paris Agreement Goal* (London: Institute for Public Policy Research and Chatham House, 2023).

13. UN, "Secretary-General's Opening Remarks to the UN Global Compact Leaders Summit," June 15, 2021.

14. Peter Newell, Shilpi Srivastava, Lars Otto Naess, Gerardo A. Torrest, and Roz Price, "Toward Transformative Climate Justice: Key Challenges and Future Directions for Research," *Institute of Development Studies Working Papers 2020* 540 (2020).

15. Jessica Kyle, "Transformational Change Is Needed to Address Climate Change" (ICF Climate Center, July 15, 2021).

16. Karen H. Olsen and Neelam Singh, eds., *Transformational Change Methodology: Assessing the Transformational Impacts of Policies and Actions* (Washington, DC: World Resources Institute, 2020).

17. Mitigation Action Facility, "Transformational Change," n.d., https://mitigation-action.org.

18. Jo-Ellen Parry, *Building a Climate-Resilient City: Transformational Adaptation* (International Institute for Sustainable Development [Winnepeg: IISD] and the University of Winnipeg, 2017).

19. Christoph Görg, Ulrich Brand, Helmut Haberl, Diana Hummel, Thomas Jahn, and Stefan Liehr, "Challenges for Social-Ecological Transformations: Contributions from Social and Political Ecology," *Sustainability* 9, no. 7: 1045 (2017), DOI 10.3390/su9071045.

20. Lauren Rickards and Stuart Howden, "Transformational Adaptation: Agriculture and Climate Change," *Crop and Pasture Science* 63, no. 3 (2012): 240–250; Matthew J. Colloff, Berta Martín-López, Sandra Lavorel, Bruno Locatelli, Russell Gorddard, Pierre-Yves Longaretti, and Gretchen Walters et al., "An Integrative Research Framework for Enabling Transformative Adaptation," *Environmental Science and Policy* 68 (2017): 87–96.

21. Karen O'Brien, Linda Sygna, Asuncion Lera St. Clair, Per Olsson, Heide Hackmann, and Eva Bakkeslett, "The First Transformation in a Changing Climate Conference: Introduction and Reflections," in *University of Oslo, Proceedings:*

Transformation in a Changing Climate, International Conference (June 19–21) edited by Karen O'Brien, Linda Sygna, Asuncion Lera St. Clair, Per Olsson, Heide Hackmann, and Eva Bakkeslett (Oslo: University of Oslo, 2013), p. 4.

22. Karen O'Brien and Linda Sygna, "Responding to climate change: The three spheres of transformation," in *University of Oslo, Proceedings: Transformation in a Changing Climate, International Conference in Oslo, Norway (June 19–21),* edited by Karen O'Brien, Linda Sygna, Asuncion Lera St. Clair, Per Olsson, Heide Hackmann, and Eva Bakkeslett (Oslo: University of Oslo, 2013), p. 16.

23. Johanna Mustelin and John Handmer, "Triggering Transformation: Managing Resilience or Invoking Real Change?" in *University of Oslo, Proceedings: Transformation in a Changing Climate, International Conference in Oslo, Norway (June 19–21),* edited by Karen O'Brien, Linda Sygna, Asuncion Lera St. Clair, Per Olsson, Heide Hackmann, and Eva Bakkeslett (Oslo: University of Oslo, 2013), p. 25.

24. Thomas M. Tanner and Aditya V. Bahadur, "Distilling the Characteristics of Transformational Change in a Changing Climate," in *University of Oslo, Proceedings: Transformation in a Changing Climate, International Conference in Oslo, Norway (June 19–21),* edited by Karen O'Brien, Linda Sygna, Asuncion Lera St. Clair, Per Olsson, Heide Hackmann, and Eva Bakkeslett (Oslo: University of Oslo, 2013), p. 34.

25. IUCN (International Union for Conservation of Nature), *Nature 2030: One Nature, One Future* (Marseilles: IUCN World Conservation Congress, 2021).

26. World Wildlife Fund, "Climate," n.d., https://www.worldwildlife.org.

27. Sophie Boehm, Katie Lebling, Kelly Levin, Hanna Fekete, Joel Jaeger, Richard Waite, and Anna Nilsson et al., *State of Climate Action 2021: Systems Transformations Required to Limit Global Warming to 1.5°C* (Washington, DC: World Resources Institute, 2021).

28. Karen O'Brien, "Global Environmental Change II: From Adaptation to Deliberate Transformation," *Progress in Human Geography* 36, no. 5 (2012): 667–676.

29. Giuseppe Feola, "Societal Transformation in Response to Global Environmental Change: A Review of Emerging Concepts," *Ambio* 44, no. 5 (2015): 376–390.

30. Sarah Park, Nadine Marshall, Emma Jakku, Anne-Maree Dowd, Stuart Howden, Emily Mendham, and Aysha Fleming, "Informing Adaptation Responses to Climate Change Through Theories of Transformation," *Global Environmental Change* 22, no. 1 (2012): 115–126.

31. Frank Biermann, Kenneth Abbott, Steinar Andresen, Karen Bäckstrand, Steven Bernstein, Michelle M. Betsill, and Harriet Bulkeley et al., "Navigating the Anthropocene: Improving Earth System Governance," *Science* 335, no. 6074 (2012): 1306–1307.

32. Brian C. Chaffin, Hannah Gosnell, and Barbara A. Cosens, "A Decade of Adaptive Governance Scholarship: Synthesis and Future Directions," *Ecology and Society* 19, no. 3 (2014), p. 56.

33. Carl Folke, Stephen R. Carpenter, Brian Walker, Martin Scheffer, Terry Chapin, and Johan Rockstrom, "Resilience Thinking: Integrating Resilience, Adaptability and Transformability." *Ecology and Society* 15, no. 4 (2010), p. 20.

34. Karen O'Brien, "Global Environmental Change II: From Adaptation to Deliberate Transformation," *Progress in Human Geography* 36, no. 5 (2012), p. 668.

35. Robert W. Kates, William R. Travis, and Thomas J. Wilbanks, "Transformational Adaptation When Incremental Adaptations to Climate Change Are Insufficient," *Proceedings of the National Academy of Sciences of the United States of America* 109, no. 19 (2012): 7156–7161.

36. Samuel Wilson, Leonie J. Pearson, Yoshihisa Kashima, Dean Lusher, and Craig Pearson, "Separating Adaptive Maintenance (Resilience) and Transformative Capacity of Social-Ecological System," *Ecology & Society* 18, no. 1 (2013), p. 22.

37. Michele-Lee Moore, Ola Tjornbo, Elin Enfors, Corrie Knapp, Jennifer Hodbad, Jacopo A. Baggio, Albert Norstrom, Per Olsson, and Duan Biggs, "Studying the Complexity of Change: Toward an Analytical Framework for Understanding Deliberate Social-Ecological Transformations," *Ecology and Society* 19, no. 4 (2014), p. 54.

38. Clive A. McAlpine, Leonie M. Seabrook, Justin G. Ryan, Brian J. Feeney, William J. Ripple, Anne H. Ehrlich, and Paul R. Ehrlich, "Transformational Change: Creating a Safe Operating Space for Humanity," *Ecology and Society* 20, no. 1 (2015), p. 56.

39. Ines S. Campos, Filipe M. Alves, Joao Dinis, Monica Truninger, Andre Vizinho, and Gil Penha-Lopes, "Climate Adaptation, Transitions, and Socially Innovative Action-Research Approaches," *Ecology and Society* 21, no. 1 (2016), p. 13.

40. Mark Pelling, *Adaptation to Climate Change: From Resilience to Transformation* (London: Routledge, 2011), p. 57.

41. Mark Pelling, *Adaptation to Climate Change: From Resilience to Transformation* (London: Routledge, 2011), p. 58.

42. Mark Pelling, *Adaptation to Climate Change: From Resilience to Transformation* (London: Routledge, 2011).

43. Petra Tschakert and Asuncion Lera St. Clair, "Conditions for Transformative Change: The Role of Responsibility, Care, and Place-Making in Climate Change Research," in *University of Oslo, Proceedings: Transformation in a Changing Climate, International Conference in Oslo, Norway (19–21 June 2013),* edited by Karen O'Brien, Linda Sygna, Asuncion Lera St. Clair, Per Olsson, Heide Hackmann, and Eva Bakkeslett (Oslo: University of Oslo, 2013): 267–275.

44. Petra Tschakert and Asuncion Lera St. Clair, "Conditions for Transformative Change: The Role of Responsibility, Care, and Place-Making in Climate Change Research," in *University of Oslo, Proceedings: Transformation in a Changing Climate, International Conference in Oslo, Norway (19–21 June 2013)* (Oslo: University of Oslo, 2013), p. 274.

45. Alexander Wendt, *Social Theory of International Politics* (Cambridge: Cambridge University Press, 1999), p. 315.

46. Leslie Sklair, "World Revolution or Socialism, Community by Community, in the Anthropocene?" *Journal of World-Systems Research* 25, no. 2 (2019): 297–305.

47. Albert V. Norstrom, Astrid Dannenberg, Geoff McCarney, Manjana Milkoreit, Florian Diekert, Gustav Engstrom, and Ram Fishman et al., "Three Necessary Conditions for Establishing Effective Sustainable Development Goals in the Anthropocene," *Ecology and Society* 19, no. 3 (2014), p. 8.

48. Giuseppe Feola, "Societal Transformation in Response to Global Environmental Change: A Review of Emerging Concepts," *Ambio* 44, no. 5 (2015), p. 378.

49. Giuseppe Feola, "Societal Transformation in Response to Global Environmental Change: A Review of Emerging Concepts," *Ambio* 44, no. 5 (2015), p. 381.

50. For an exception, see Ian Scoones, Andrew Stirling, Dinesh Abrol, Joanes Atela, Lakshmi Charli-Joseph, Hallie Eakin, and Adrian Ely et al., "Transformations to Sustainability: Combining Structural, Systemic and Enabling Approaches," *Current Opinion in Environmental Sustainability* 42 (2020): 65–75.

51. Ross Gillard, Andrew Gouldson, Jouni Paavola, and James Van Alstine, "Transformational Responses to Climate Change: Beyond a Systems Perspective of

Social Change in Mitigation and Adaptation," *Wiley Interdisciplinary Reviews: Climate Change* 7, no. 2 (2016): 251–265.

52. IPCC, "Point of Departure and Key Concepts," in *Climate Change 2022: Impacts, Adaptation and Vulnerability. Contribution of Working Group II to the Sixth Assessment Report of the Intergovernmental Panel on Climate Change* (Cambridge: Cambridge University Press, 2022), p. 171.

53. See Rob Roggema, Tim Vermeend, and Andy Van den Dobbelsteen, "Incremental Change, Transition or Transformation? Optimising Change Pathways for Climate Adaptation in Spatial Planning," *Sustainability* 4, no. 10 (2012): 2525–2549.

54. IPCC, "Summary for Policymakers," in *Climate Change 2023: Synthesis Report. A Report of the Intergovernmental Panel on Climate Change. Contribution of Working Groups I, II and III to the Sixth Assessment Report of the Intergovernmental Panel on Climate Change,* edited by H. Lee and J. Romero (Geneva: IPCC, 2023), p. 31, emphasis added.

55. IPCC, "Summary for Policymakers," in *Climate Change 2023: Synthesis Report. A Report of the Intergovernmental Panel on Climate Change. Contribution of Working Groups I, II and III to the Sixth Assessment Report of the Intergovernmental Panel on Climate Change* (Geneva: IPCC, 2023), p. 31, emphasis added.

56. IPCC, "Climate Resilient Development Pathways," in *Climate Change 2022: Impacts, Adaptation and Vulnerability, Contribution of Working Group II to the Sixth Assessment Report of the Intergovernmental Panel on Climate Change* (Cambridge: Cambridge University Press, 2022), pp. 2655–2807.

57. IPCC, "Climate Resilient Development Pathways," in *Climate Change 2022: Impacts, Adaptation and Vulnerability, Contribution of Working Group II to the Sixth Assessment Report of the Intergovernmental Panel on Climate Change* (Cambridge: Cambridge University Press, 2022), p. 2657.

58. IPCC, "Climate Resilient Development Pathways," in *Climate Change 2022: Impacts, Adaptation and Vulnerability, Contribution of Working Group II to the Sixth Assessment Report of the Intergovernmental Panel on Climate Change* (Cambridge: Cambridge University Press, 2022), p. 2668.

59. IPCC, "Summary for Policymakers," in *Climate Change 2023: Synthesis Report. A Report of the Intergovernmental Panel on Climate Change. Contribution of Working Groups I, II and III to the Sixth Assessment Report of the Intergovernmental Panel on Climate Change*, 1–34 (Geneva: IPCC, 2023), p. 31.

60. Roger Pielke Jr., "The Political Agenda of the IPCC," *The Honest Broker,* Substack, May 15, 2023, rogerpielkejr.substack.com/p/the-political-agenda-of-the-ipcc.

61. IPCC, "Summary for Policymakers," in *Climate Change 2023: Synthesis Report. A Report of the Intergovernmental Panel on Climate Change. Contribution of Working Groups I, II and III to the Sixth Assessment Report of the Intergovernmental Panel on Climate Change*, 1–34 (Geneva: IPCC, 2023), p. 24.

62. IPCC, "Point of Departure and Key Concepts," in *Climate Change 2022: Impacts, Adaptation and Vulnerability, Contribution of Working Group II to the Sixth Assessment Report of the Intergovernmental Panel on Climate Change* (Cambridge: Cambridge University Press, 2022), p. 172.

63. Global Commission on the Economy and Climate, *The New Climate Economy* (Washington, DC: World Resources Institute, 2018), p. 8.

64. Kate Raworth, *A Safe and Just Space for Humanity: Can We Live Within the Doughnut?* (Oxford: Oxfam, 2012).

65. IPCC, "Point of Departure and Key Concepts," in *Climate Change 2022: Impacts, Adaptation and Vulnerability, Contribution of Working Group II to the*

Sixth Assessment Report of the Intergovernmental Panel on Climate Change (Cambridge: Cambridge University Press, 2022), p. 173.

66. IPCC, "Summary for Policymakers," in *Climate Change 2023: Synthesis Report. A Report of the Intergovernmental Panel on Climate Change. Contribution of Working Groups I, II and III to the Sixth Assessment Report of the Intergovernmental Panel on Climate Change*, 1–34 (Geneva: IPCC, 2023), p. 2737.

67. Peter Newell, "The Elephant in the Room: Capitalism and Global Environmental Change," *Global Environmental Change* 21, no. 1 (2011): 4–6.

68. IPCC, "Decision-Making Options for Managing Risk," in *Climate Change 2022: Impacts, Adaptation and Vulnerability. Contribution of Working Group II to the Sixth Assessment Report of the Intergovernmental Panel on Climate Change* (Cambridge: Cambridge University Press, 2022): 2539–2654.

69. IPCC, "Decision-Making Options for Managing Risk," in *Climate Change 2022: Impacts, Adaptation and Vulnerability. Contribution of Working Group II to the Sixth Assessment Report of the Intergovernmental Panel on Climate Change* (Cambridge: Cambridge University Press, 2022), p. 2542.

70. IPCC, "Decision-Making Options for Managing Risk," in *Climate Change 2022: Impacts, Adaptation and Vulnerability. Contribution of Working Group II to the Sixth Assessment Report of the Intergovernmental Panel on Climate Change* (Cambridge: Cambridge University Press, 2022), p. 2542.

71. IPCC, "Decision-Making Options for Managing Risk," in *Climate Change 2022: Impacts, Adaptation and Vulnerability. Contribution of Working Group II to the Sixth Assessment Report of the Intergovernmental Panel on Climate Change* (Cambridge: Cambridge University Press, 2022), p. 2547.

72. IPCC, "Decision-Making Options for Managing Risk," in *Climate Change 2022: Impacts, Adaptation and Vulnerability. Contribution of Working Group II to the Sixth Assessment Report of the Intergovernmental Panel on Climate Change,* 2539–2654 (Cambridge: Cambridge University Press, 2022) p. 2551.

73. IPCC, "Summary for Policymakers," in *Climate Change 2022: Impacts, Adaptation and Vulnerability. Contribution of Working Group II to the Sixth Assessment Report of the Intergovernmental Panel on Climate Change* (Cambridge: Cambridge University Press, 2022), p. 12.

74. IPCC, "Summary for Policymakers," in *Climate Change 2022: Impacts, Adaptation and Vulnerability. Contribution of Working Group II to the Sixth Assessment Report of the Intergovernmental Panel on Climate Change* (Cambridge: Cambridge University Press, 2022), p. 20.

75. IPCC, "Summary for Policymakers," in *Climate Change 2022: Impacts, Adaptation and Vulnerability. Contribution of Working Group II to the Sixth Assessment Report of the Intergovernmental Panel on Climate Change* (Cambridge: Cambridge University Press, 2022), p. 25.

76. IPCC, "Summary for Policymakers," in *Climate Change 2022: Impacts, Adaptation and Vulnerability. Contribution of Working Group II to the Sixth Assessment Report of the Intergovernmental Panel on Climate Change* (Cambridge: Cambridge University Press, 2022), p. 26.

5

Adaptation and Capitalism

THE DISCUSSION IN THE PRECEDING CHAPTERS HAS engaged an analysis of the idea of "transformation" in climate policy. The record shows that adaptation was on the scientists' agenda from the beginning and has become increasingly significant for policymakers as the severity of the climate crisis has been widely accepted. To some extent, adaptation has been advanced as an alternative to mitigation, thus serving the interests of fossil fuel companies and major oil-, gas-, and coal-producing countries. However, the main advocates of stronger support for adaptation in international negotiations have been developing countries facing the most severe effects of climate change and lacking the resources to respond effectively to the impacts of a much warmer world. Most developing countries want significant mitigation along with ample funding for adaptation to the now-unavoidable harms resulting from climate change. Adaptation has risen steadily on the global agenda, resulting in the creation of several funds and entities devoted to supporting developing country adaptation efforts. Scholars, and with them the Intergovernmental Panel on Climate Change (IPCC), have moved toward the topic because the science regarding global warming has reached a settled consensus, and the question for them has become what to do about it. Meanwhile, policymakers have turned to adaptation as the failures of mitigation have become obvious.

At the Threshold of Transformative Change

The adaptation policy dialogue has dealt with several abiding issues: finances, finding a balance between mitigation and adaptation, historic responsibility for the damages resulting from climate change, and the role of the public and private sectors in adaptation. Along the way, the principle of common but differentiated responsibilities has largely been lost, as shown recently in the reframing of loss and damage as an aspect of adaptation rather than as a legal claim of liability and compensation for damages done.[1] More broadly, at every turn in the adaptation discourse that the response to climate change could have meant transformational change in global, national, and local institutions, adaptation has instead been folded into the traditional development agenda, clearly revealed in the IPCC and UN linking of adaptation to the Sustainable Development Goals.

Meanwhile, policymakers, advocates, and scholars, including the UN Secretary-General, insist that climate risks call for global transformation, but with little explanation of what that would mean in practical terms. Rather than significant change in global institutions, transformation is often framed as cognitive change toward ecological attitudes and beliefs, coupled with a general call for sustainability and renewable energy, but with little guidance on what institutional changes are expected to follow from cognitive reorientation. All of this leaves us with the critical question: What is to be done? Johanna Wolf, Karen O'Brien, and Linda Sygna write, "In light of the social and environmental transformations that current greenhouse gas emissions trajectories imply, the ultimate question may not be whether or not transformations are indeed coming, but rather whether and to what extent humanity will mobilize its collective agency to actively participate in, rather than passively endure, transformations."[2] The world will change, but human society might or might not be prepared for what is to come.

The main topic of this chapter and the one that follows is a review of alternative visions of adaptation and transformation forwarded in climate advocacy and scholarship. Most are concerned with saving civilization as we know it, encompassing varying levels of ambition regarding the institutional changes required to do so. Ecological modernization, one of the two main options presented in the most recent IPCC Working Group II report, poses a serious and widely accepted view of what is required, one that does not disturb existing economic institutions. Against this orientation, and its celebration of market-led economic growth as the solution to ecological problems, are the steady state, degrowth, ecosocialism, and ecological civilization. For them, "Breaching adaptation limits will result in escalating losses or require transformational change."[3] There is also a middle ground that includes green growth and calls for significant change in organizations such as the United Nations while still building on existing socioeconomic arrangements.[4]

Meanwhile, other voices in adaptation discourse, to be assessed in Chapter 7, have given up on civilization as we know it altogether, believing that it is too late for modern society to accept and implement the sweeping changes necessary to avert climate catastrophe. For them, the future portends a tough new planet, or a global Dark Age, about which the best we can do is some form of Deep Adaptation to preserve the human species while the institutions of today's global order collapse. Transformational adaptation and change will occur by necessity, as "imposed adaptation"; that is, humans will be coping as well as possible with a global cataclysm that will sweep away modern society and probably most of the world's human population as well.

Ecological Modernization and the Growth Imperative

Ecological modernization theory (EMT) emphasizes the role of technological innovation driven by the profit motive in responding to ecological challenges.[5] This is especially important for adaptation, as many observers and policymakers argue that alleviating poverty through growth and disseminating appropriate technologies constitute the best response to climate risks. John Bellamy Foster observes, "Ecological contradictions, according to this ideology, can be surmounted by means of technological fixes and continued rapid growth in production, with no fundamental changes to the structure of our economy or society."[6] Thus, ecological modernization theory represents a defense of modern society, in which *modern* is understood as the combination of scientific, technological and industrial progress with the greater extent of the globalized, integrated market and the drive for economic growth.[7]

Advocates of the theory see modern society as adaptable and capable of adopting sustainable policies and practices and, even further, that modernization offers the best way to achieve sustainability. Market incentives, free trade, relatively easy immigration, and technological upgrades, such as geoengineering[8] and, more prosaically, installing air conditioning,[9] can ensure adequate adaptation to climate change while excessive government interference in markets will thwart effective adaptation.[10] Political change is expected to accompany economic change, so the structure and agencies of government will support economic shifts toward efficiency. In short, rather than reining in industrial development and economic growth, the theory sees them as key to resolving global ecological problems.

According to EMT advocates, the profit motive and entrepreneurship can enable the development and spread of technologies and practices that better protect the environment and provide the means for effective adaptation. Many measures to improve efficiency and thus profitability would also reduce the human impact on the planet. Further, prices for the goods needed to protect a household against climate impacts steadily decline, bringing such items as air conditioning within reach of billions of people.

More broadly, contrary to the supposed limits to growth, EMT considers economic growth to be essential to sustainability.[11] Evidence for this claim can be found in the poor environmental records of less-developed countries and socialist states compared to the environmental improvements witnessed in wealthy, industrially developed countries. The more modern a country is, the more it protects the environment. Indeed, modern institutions such as corporations and states will see protecting the environment to be in their own self-interest, thus implementing "widespread ecological reforms, without requiring radical social or political-economic change."[12] While not drawing its own conclusions as to which is preferred, the IPCC suggests that EMT and degrowth are the two main alternatives available. In practice, the political leaders of the world, in line with powerful economic interests, have opted for EMT, even while advocating transformation.[13]

Critics of this view note that the empirical support for it is weak: modern society has shown no greater capacity to manage ecological challenges than non-modern societies, reforms intended to protect the environment can be more symbolic than real, and examples of success in particular cases do not disprove that the broader tendency of modernity is toward ecological degradation.[14] As Jeff Goodell remarks, adaptation has its limits, "and the notion that eight billion people are going to thrive on a hotter planet by simply cranking up the air-conditioning or seeking refuge under a pine tree is a profound misunderstanding of the future we are creating for ourselves."[15]

EMT and its cognates rely on the notion that everything significant can be evaluated in monetary terms. Against this, Carlo Fanelli writes, "Since orthodox economists assume that everything has a price and that the market will inherently sort out all problems, mainstream approaches to climate change suffer from an inherent inability to deal adequately with the climate crisis."[16] Further, in practice, growth outstrips efficiency gains, so the overall impact of industrial production increases even as it becomes more efficient. Most tellingly, EMT ignores that richer countries can displace, rather than eliminate, their environmental impacts onto poorer countries, where exploitative resource extraction and polluting industries degrade the environment to serve rich-country markets.[17] Thus, the appearance of ecological progress in a given country obscures the reality of greater ecological damage occurring on a global scale. Contrary to EMT, asserts Foster, "Any ecological defenses erected in the present must be based on opposition to the logic of capital accumulation."[18]

The Growth Imperative and Adaptation

The debate on alternative futures for environmental and climate policy centers on the question of economic growth. Scholars, advocates, and policymakers differ on whether economic growth is ecologically viable or beneficial. They ask whether different types of growth exist, with some recommending "green growth" that employs market incentives and public investment to achieve decarbonization and other pro-environment policies. Others argue that growth is the problem, and it is time to halt and even reverse growth so humans have less impact on the global ecology, including its climate.[19] In turn, the question arises as to whether growth is inherent and required in a capitalist society. Matthew Paterson succinctly poses the problem: "But raising the question of growth also raises the question of capitalism, since the most common explanation within political economy (especially but not only ones derived from Marx) frameworks for the question of why we live in a world that is obsessed with growth, is because that world is capitalist."[20] Regarding adaptation to climate change, this places at the forefront whether capitalism must be replaced with a different social order for sufficient adaptation to be possible.

The stakes in the debate over growth could not be higher. Although considerable progress has occurred over the past few decades, approximately 700 million people remain in deep poverty worldwide, living on less than $2.15 per day. Presumably, economic growth is the remedy for poverty, and few would argue that the dire circumstances of the poor should go unaddressed to save the environment while others continue to enjoy comfortable lifestyles. Yet unchecked, market-led economic growth, some say, would have the result of driving the world into ecological catastrophe, meaning even greater suffering in the future. A choice between tolerating deep poverty for hundreds of millions on one hand, or facing ecological collapse on the other, is not one the world wants to face. If successful adaptation to climate change requires slowing, halting, or even reversing economic growth, then what is to be done about the growing number of poor people and those living in precarious circumstances?

Growth and Ecological Modernization Theory

As noted, EMT solves this dilemma by positing that economic growth is the solution, not the problem. This directly opposes the "limits to growth" argument put forth in 1972,[21] along with arguments that population growth was leading the world toward ecological crisis. Regarding climate change, a recent article in line with EMT suggests that adaptation will take care of any problems that might arise from further burning of fossil fuels. Supposedly, evidence that the level of damage from extreme storms has actually declined shows that adaptation is feasible and will allow business as usual. According to this view, wealthier and more developed countries have the means to implement appropriate disaster warning and protection against extreme weather events, so the answer is to ensure that all countries have sufficient development to meet the challenge as it arises. Growth is the solution to the climate problem.[22]

Against this, one finds analysts who believe that it is impossible to grow indefinitely on a finite world, and many of today's political conflicts reflect emerging scarcities and limits that will only worsen with time. Paul R. Ehrlich and Dennis C. Pirages note the arrival of many linked crises, including "climate disruption, water shortages, shrinking energy and mineral supplies, the loss of biodiversity and essential ecosystem services, toxification of the planet, and the growing threat of pandemics."[23] As humanity's global ecological footprint increases, these problems worsen. In short, a possible crisis in world affairs could result from overpopulation, overconsumption, heedless exploitation of natural resources, environmentally harmful technologies, and institutions geared toward maximizing economic growth.[24]

Although many issues remain controversial, in this the EMT advocates are unpersuasive. For one, the observed and potential harms from climate change do not reduce to one risk and one measure of successful adaptation, as Ted Nordhaus, Vijaya Ramachandran, and Patrick Brown suggest.[25] Even if the loss of human life from extreme weather events reduced to zero, nothing would have been done about species loss, declines in agricultural productivity, heat-related illnesses, and a wide range of other challenges outlined in IPCC reports and elsewhere. The limits to growth analysts do far better at capturing the multifaceted character of the climate problem. Moreover, as Laurie Parsons correctly observes, ongoing ecological destruction in the poorer countries of the world might occur because of global economic development, not in spite of it. As EMT would have it, advanced countries have progressed and evolved past the stage of heedless ecological exploitation, and the rest of the world can and will do so as well. By contrast, Parsons asks "What if one place was devastated *because* the other was clean? What if one place was at risk *because* the other was safe?"[26] Moving the location of ecological degradation is obviously no solution to the global problem of climate change. Indeed, regarding adaptation, the upshot would be exacerbation of what has been widely discussed for decades: the most pressing adaptation needs appear in those places lacking the resources to reduce vulnerability and enhance resilience.

It may well be that wealthier countries are less prone to dumping toxic wastes in their waterways today than they were decades ago, or compared to poorer countries today. Even so, the fact remains that economic growth in the contemporary global economy means more burning of fossil fuels. The amount of fossil fuels consumed rises steadily, only temporarily pausing when a recession or a global pandemic slows the economy. To be sure, the share of energy provided by renewable sources continues to increase, but that does not translate into a reduction of fossil fuel use; both rise together. Consequently, growth as the means by which countries are to acquire the means to adapt successfully to a warming world is self-defeating. The greater adaptive capacity resulting from growth must cope with higher temperatures and climate impacts also resulting from growth.

Green Growth

Green growth answers the critique of EMT by arguing that growth can be achieved while society adopts consciously pro-environment policies. The theory is similar to EMT, although with perhaps a stronger hand for governments in guiding economic activity in the right direction. Green growth advocates suggest that growth need not require a larger ecological footprint. To the contrary, the shift to services, entertainment, and the virtual world mean that consumers demand, and businesses supply, products with light ecological impacts. In addition, with proper incentives, economic activity can be made far more efficient, producing as much and more with less energy. Regarding climate, this path rests on the feasibility of "decoupling" growth from the burning of fossil fuels.

Like EMT, governments and international organizations have embraced the notion of green growth, and institutes to advance the program disseminate supporting research and information. Accordingly, analysts issue plans for enhancing the capacity of international organizations to provide governance toward green growth. For example, Mark Harvey suggests that the United Nations be granted the power to tax so as to create incentives for decarbonization and to fund climate transitions. He writes, "Provocative perhaps, but the UN needs supranational tax-raising powers for a progressive tax on national wealth, both to reduce the CO_2 emissions of the wealthiest and promote the green development of the less wealthy economies. In short, to promote planetary egalitarianism over the planet's resources and sustainability."[27] However, one would think that if the members of the UN can agree to supranational taxing authority, they could instead agree to binding emissions targets and adequate funding for adaptation efforts, avoiding the encroachment on national sovereignty implied in the tax proposal. In any event, reformed international organizations are an element of the green growth program.

The attraction of green growth is that it offers a remedy for the entire range of environmental problems, including mitigating and adapting to climate change, that seems to be within reach of existing institutions and that does not challenge the strong tendency, if not imperative, for the global economy to expand. Consequently, economic development can proceed, continuing to be the means by which deep poverty is to be eliminated without need for a massive redistribution of global wealth or restructuring of class relations.

Yet critics find the evidence fails to support green growth. Decoupling, they say, does not provide enough gains to offset the higher resource use implied by economic growth.[28] Moreover, the transition from the current industrial economy to a green economy cannot occur fast enough to avert significant environmental harm, including surpassing the 1.5°C and 2.0°C climate thresholds.[29]

That said, adaptation, taken by itself, could fare well under a broad program of green growth. For example, projects aimed at enhancing resilience and reducing vulnerability might add to gross domestic product (GDP) while also reducing harms from climate change. A combination of public and private finance could fund strengthening infrastructure, enhanced building design, new methods of farming, and more. The profit motive and normal public policy to provide collective goods would be compatible with green growth, with no need for sweeping institutional change. Thus, adaptation *is* growth.

The question would be whether green growth is adequate to the task. Adaptation as part of a green growth program could fall ever further behind the worsening problem, perhaps even adding to environmental pressures with more construction of adaptation projects or more widespread installation of air conditioning.[30] Also, instead of solving the problem of adapting to climate change, growth-oriented adaptation could simply delay the inevitable reckoning for a few years. The delay could make matters worse if the result is an even bigger crisis ensuing as the climate dam breaks.

Does Capitalism Require Growth?

Green growth and EMT preserve the premise of a growing economy. But does capitalism require growth at all? If so, then unless the highest hopes for decoupling are fulfilled, the system will inevitably run up against the problem of infinite exponential growth on a finite planet. A capitalist world order facing limits on the growth imperative, leading to economic and social crisis, might not be able to opt for a long-term plan for adaptation to climate change that is sustainable, equitable, and effective. Consequently, any incremental efforts in the short or medium term to ensure perpetuation of a capitalist order would represent maladaptation; only those measures that lead toward transformational change could properly be called *adaptation*. If capitalism does not require growth, or it can embrace some form of green growth, then a stable, sustainable level of resource throughput can be achieved while maintaining institutions nearly as they are. But numerous studies and analyses argue that this is not possible.

According to some observers, capitalism that is not growing is a contradiction in terms. The system is based on profits and the accumulation of capital. Accumulation means a growing mass of productive assets, from mines to factories to transportation systems, and all of that requires more resources and more energy. Profits exist to be reinvested so as to grow the enterprise and, as all enterprises compete to become larger, the entire economy grows as well. Any interruption of this process is a crisis, the severity of which can vary from a mild recession to setting the stage for world war. If capitalism cannot halt growth without throwing itself into a social, political, and economic crisis, some options for addressing the climate challenge simply do not exist. That is to say, capitalism itself creates hard limits to adaptation, limits that only transformation toward an entirely different society can overcome.

Several explanations for why capitalism requires growth have been forwarded. All these explanations must grapple with the fact that capitalism has endured periods of slow or no growth without collapsing. Countless recessions and a number of major depressions mark the history of global capitalism. Most appear to be temporary manifestations of the business cycle, while others seem to arise from deep attributes of the system. But capitalism survives, more often than not roaring back in a boom of growth, prosperity, rising consumption, and full employment. Why should this pattern not continue forever? Why have the periods of doldrums not so far spelled the end of the system?

One explanation for the growth imperative is that capitalism is always facing crises of overproduction and underconsumption. Enterprises collectively employ the available labor to produce commodities for sale on the market, but they do not return to the workers enough in wages to buy all that is produced. If they did, there would be no profits. Thus, some goods go unsold, and prices must be slashed, wiping out the profit margin. With no profit expectations, investment dries up, unemployment increases, consumers have less purchasing power, and a downward spiral toward depression begins. In the modern world, government deficits, defense spending, and borrowing to fund government and consumption fill the gap. Thus, recessions are relatively brief and do not become depressions. However, each downturn becomes more serious, and the government interventions required to stave off a depression become larger, such that the fiscal health of the government comes into question. Eventually, the stopgaps to save the system from the paradox of glut alongside poverty will no longer work, and a systemic crisis will arrive again, just as one did in 1929. Avoiding the crisis of overproduction and underconsumption requires a growing economy, one that is always absorbing surplus goods and offering attractive investment opportunities but, critics say, the dynamic cannot go on forever.

When things are bad enough, society faces a difficult choice. On one hand, the cycles of boom and depression cannot continue forever because opponents to a system that so often threatens well-being and even survival will resist and overthrow it, establishing an alternative social order without capitalism's ills. Or, on the other hand, the ever-deeper crises afflicting the system will result in dissolution of society. The ecological challenges, including global warming and incapacity to adapt to it, will only worsen a civilizational collapse akin to the fall of the Roman Empire. The choice, we are told, is between socialism or barbarism, and there is no knowing which holds the upper hand. Either way, it will be clear to everyone that continuing business as usual is no longer an option. And long before this moment comes, all measures to sustain the rate of growth will be employed, meaning that sooner or later production limits intended to protect the global atmospheric commons will be cast off, and public funds devoted to adaptation will be diverted to boosting profits.

Another take on the question locates the growth imperative at the firm level rather than in broad systemic phenomena such as underconsumption and overproduction. Capitalism as a system might be compatible with slow growth or a steady state. But firms cannot tolerate such conditions, and their individual demands for expansion add up to a systemwide growth imperative. David Schweikert writes that every capitalist firm wants to grow without limit out of greed and fear. The owners are in it to become wealthy, an insatiable demand, and failing to grow the firm risks being eliminated from the game by other companies.[31] By this account, capitalism does not require growth as such; it requires that investors and firms perceive profit opportunities and so choose to invest. Without this dynamic, the economy can survive, but it is not healthy, and an unhealthy economy is bad for investors, workers, the government, and the environment.

Regardless of which mechanism explains the growth imperative, for capitalism as currently constituted, an interruption in growth (as measured by GDP) does create a social crisis. Leaders in government and industry know this, and they bend every effort toward averting interruptions, or even slowdowns, of growth, and making those that do occur as short and harmless as possible. Any restraints in place for environmental protection are quickly discarded if necessary to ensure resumption of a rising level of economic activity. Workers are as tied to this imperative as capital because they want the jobs and decent pay attending a growing economy, and to avoid the dire personal consequences of economic downturns. Politicians in all countries heed the growth imperative, knowing they are at risk of losing the next election or being overthrown if they do not maintain the conditions for capital accumulation. In addition, an entire industry has emerged to encourage consumption so that markets for goods and services are robust and growing, ensuring more opportunities for investment, reaping profits, and reinvesting the gains in ever larger firms and markets. The treadmill of production spins faster and there is no way off of it within existing arrangements.

Ecosocialists and other critics assert that ecologically damaging growth is inherent in capitalism, constituting another major contradiction alongside intensifying class conflict.[32] In the economic context of a slump, investing in adaptation and other forms of environmental protection will likely fall down the list of priorities. A government facing declining revenues will more likely prioritize measures to restore business investment and perhaps to protect the neediest members of society rather than proactive expenditures to address a problem still some years in the future. Possibly, in some cases adaptation to climate change will appear to thwart growth objectives, in others to be irrelevant and low priority.

The challenge that global growth presents for mitigation is straightforward: more growth means more burning of fossil fuels and other additions to the factors causing global warming. If no-growth capitalism is a contradiction, then capitalism is incompatible with mitigation, and transformation of the world economic order will be required to avert rising global temperatures. However, the relationship between growth and adaptation is not as obvious. Indeed, as we have seen, some observers believe that economic growth contributes to adaptive capacity. This requires closer scrutiny of the proposition that a growth-oriented capitalist system would thwart successful adaptation to climate change.

Capitalism and Adaptation to Climate Change

Why would capitalism do any better or worse than any other social system at adapting to severe climate change? The claim that a better alternative exists is comparative and, as such, requires elaboration. Joel Kovel aptly frames the problem, noting that the central question is whether capital "can adapt and change its ways, given the gathering breakdown of its natural ground—or to be more exact, whether it can do so in time to permit mending of its relationship with nature."[33] One must explain, not simply assert, why adaptation in a global capitalist system would fail. Given the potentially catastrophic effects of 3°C or more temperature rise, it is also necessary to ask why a society built on different foundations would have the attributes needed to maintain civilization as we know it while capitalism could not. After all, perhaps no social, economic, and political order can do so.

Regarding the ecological crisis generally, and climate change mitigation in particular, many analysts have elaborated on why global capitalism has and will continue to march toward eco-catastrophe.[34] Leftist researchers point to capitalism's growth imperative, the limits to growth, its exploitation of nature, and the entrenched interests that thwart rational and feasible measures to move to renewable energy sources. They agree with Chris Williams that "ecological devastation is not an accidental outcome of capitalist development but an intrinsic element of the system, just as integral as class exploitation, poverty, racism, and war."[35] As noted, from such analysis solutions follow fairly directly. If global capitalism stands in the way of mitigating global warming and averting transgression of ecological boundaries, then replacing capitalism with ecosocialism or some other greener system is the answer. While that struggle continues, popular movements must demand now that policies to reduce greenhouse gases (GHG) be enacted that will bridge the temporal gap between today's global neoliberalism and a future ecologically sound world.

However, adaptation does not pose the same challenges as mitigation, and it is not obvious that the same diagnosis and remedy apply. The broad critiques of capitalism indicate why the system is not capable of reducing human impacts on the environment. They explain why capitalism cannot reduce pollution (of which GHG emissions are one kind) or conserve natural resources. Adaptation, by contrast, is about how societies cope with the unwanted side effects of modern industrial production and high-energy ways of life. It may well be entirely true that "capitalist societies are primed to generate recurrent environmental crises throughout their history."[36] But that does not in itself imply that capitalist societies are unable to respond effectively enough to the ecological crises that capitalism creates to ensure perpetuation of the system.

Mitigation has failed because it requires a collective response, one that global capitalism has not shown that it can provide. But adaptation takes place in large part without such a collective effort. As communities, regions, and nations experience or anticipate the impacts of climate change, they will respond, even if inadequately. To the extent that they take conscious steps to reduce their vulnerability and increase their resilience, they will have at least attempted to adapt rather than simply react. Unlike mitigation, each community can choose whether and how to adapt, regardless of what others do to adapt. In the language of rational choice theory, adaptation is a problem of externalities rather than public goods.

Given the decentralized nature of the problem, it is not certain that a capitalist system would fail to adapt sufficiently to survive or that an alternative social order would do any better. Indeed, one could make the case that capitalism would have advantages regarding adaptation to climate change. For one, markets and the price mechanism could lead firms to respond quickly and flexibly to local conditions, which is appropriate to the local character of adaptation. The price signal might efficiently register new conditions from a wide range of influences, including changing requirements of living in communities coping with a warmer world. Climatic impacts will vary widely across different locations, requiring responses adapted to local conditions. One could argue that markets and prices offer the best way to lead people toward appropriate adjustments without imposing a top-down solution that may or may not enact measures best suited to conditions on the ground. Decentralized capitalist markets might offer the flexibility needed to respond to changing conditions.

To be sure, profit-driven markets can produce externalities, shifting costs to third parties who had no responsibility for creating the problem. Thus, as things stand, developing countries and poor communities in developed countries will pay much of the cost of adaptation to a global warming crisis that they did not have much hand in creating. Yet one could argue that getting the incentives right, compensating losers, and instituting effective environmental regulations would address problems without the need to transform global capitalism into a form of ecosocialism. Indeed, that is the more or less explicit view of the negotiators at the Conferences of the Parties (COPs).

Moreover, as noted in the discussion of EMT and green growth, defenders of capitalism argue that adaptation will benefit from economic growth. Regarding mitigation, growth means more energy use and thus more burning of fossil fuels, worsening the problem. By contrast, increased adaptive capacity will be essential to successful adaptation, and economic growth could provide that capacity. If the problem is that the poor will bear heavy costs because they lack the means to implement sufficient adaptation to reduce vulnerability and increase resilience, then the solution is to alleviate their poverty, thus increasing their capacity. The economic growth that capitalism generates presumably provides the wherewithal to reduce global poverty and therefore to increase resilience and adaptive capacity.

Besides, one might argue, the monumental task of overthrowing global capitalism is a multigenerational project for which time is lacking in the face of a looming climate crisis. It is better, then, to devise and implement adaptation measures within existing social, political, and economic institutions. In sum, properly incentivized markets and appropriate regulation could be sufficient to institute effective adaptation to a warmer world and, in light of the short time available, constitute the only viable option anyway.

What is wrong with this defense of capitalism's ability to adapt to rising global temperatures? To begin, the supposed strength of capitalist adaptation is, instead, a potential weakness. Relying on price signals could foreclose the advance planning that will be required to respond effectively to global temperature rise, especially as the world moves beyond the 2°C threshold. Thus, the sensitivity and flexibility that would allow fine-grained responses to local conditions can also be barriers to comprehensive, well-designed adaptation.

Capitalist markets lack sufficient capacity for anticipatory adaptation. Production and marketing decisions are made to ensure profits. In most industries, this requires firms to take a short-term orientation. To please stock markets and pad executive compensation, the time horizon might be the next quarter, not the years or decades that some forms of adaptation will require. One might argue that this is where effective regulation enters, to make sure the incentives are right and to provide the context for rational, long-term planning. Yet the record shows that industries often capture their regulators. More broadly, governments have turned toward deregulatory philosophies like neoliberalism, empowering core capitalist enterprises while disabling movements for ecological sustainability.

Further, externalities are extremely difficult to remedy, especially in the anarchic international system and in domestic societies marked by extreme power inequalities. To do so requires some way to bring about transfers from those causing the externalities to those suffering the consequences, and efficiency mandates that the full costs of environmental harms to third parties be paid. Regarding adaptation to climate change in the global context today, that means significant sums flowing from the wealthy countries to the less-developed, poor countries. Nothing suggests such transfers will occur on the scale necessary to meet the challenge, compensate those harmed, or ensure that the full costs are reflected in prices.[37] As Mike Davis observes, "In a warmer world, moreover, socio-economic inequality will have a meteorological mandate, and there will be little incentive for the rich northern hemisphere countries, whose carbon emissions have destroyed the climate equilibrium of the Holocene, to share resources for adaptation with those poor subtropical countries most vulnerable to droughts and floods"[38] (as well as, we might add, the numerous other impacts of global warming). What is more, governments committed to various forms of neoliberalism at home cannot be relied on to ensure adequate redistribution of wealth to redress externalities imposed on their own less-privileged people, much less to people in other countries.[39]

In addition, measures taken to adapt in one locale can have negative impacts on other areas.[40] For example, building dams and levees to control flooding could negatively affect agriculture downstream, perhaps in other countries. Thus, the sensitivity to local conditions that capitalism enjoys can be a liability in some instances. This would suggest some form of management is required, whether national regulations or international agreements. But that is to say that the price mechanism and individual choices are not adequate to the task of ensuring successful adaptation. How extensive institutional transformation must be to provide successful adaptation might vary from light regulatory management to fundamental transformation of social, political, and economic arrangements. The likelihood that adaptation in one country could negatively affect conditions in other countries, rendering adaptation an international concern, suggests that the needed reforms would be sweeping.

Further, in today's context of predominant finance capital, investors are able to move quickly from areas that are no longer sufficiently profitable to more lucrative locations.[41] Consequently, if climate change imposes significant costs on the production process, the preferred response might be to abandon the area and reestablish operations elsewhere. Local taxes to pay for adaptation measures will be resisted and evaded, first with the threat of relocation, then with actual relocation. Regions and communities with the greatest adaptation needs will face low investment and disinvestment.

Inequality pervades the global system and hinders prospects for equitable, effective adaptation to climate change. The increasing concentration of wealth and income in fewer hands is an abiding attribute of global capitalism. For as long as possible, implementation of adaptation will correspond to varying levels of wealth and privilege, with those at the top ensuring that their lives and property are much better protected than those at the bottom. Conceivably, as temperatures rise, civilization as currently structured will break down, leaving no one protected. Nevertheless, the privileged will not likely implement egalitarian social arrangements that could prevent catastrophic collapse. Just as in prior civilizational crises, they will cling to their privileges while the world crumbles around them.[42] Meanwhile, those in deep poverty, still about 700 million people worldwide today, will be exposed to the worst effects of the climate crisis.

Crucially, as conditions worsen, the implicit premise of today's global adaptation policy—that adaptation needs will be located primarily in developing countries, requiring voluntary contributions from the wealthy in core countries to the vulnerable in the periphery—will be shown false.[43] Adaptation will become a growing problem for the core, exceeding even rich countries' adaptive capacity, as temperatures rise. Mark Lynas writes, "Civilisational collapse, like the blast wave of a neutron bomb, will sweep around the globe."[44] Consequently, the externalities wealthy countries have imposed on distant places and the usual victims of capitalist development will come home in dramatic, undeniable ways. Indeed, as industrial civilization frays, the adaptive capacity of the capitalist core could prove to be lower than in communities and regions less reliant on high-energy, complex, integrated economic and social systems.[45] How can today's institutions respond?

As we have seen, growth is the highly problematic answer that global capitalism has offered to address the deep poverty that afflicts so many people today. But to avoid severe environmental impacts while relying on economic growth as the means to reduce poverty requires redistribution. Without significant redistribution, the world will get the growth for as long as it lasts, but not the environmental benefits. "This simultaneous need for economic growth among parts of the world population and for economic stagnation or reduction in others lies at the heart of the inequality issue and—indeed—of the sustainability challenge," Iris Borowy observes.[46] Yet the power of the owners of capital to prevent significant redistribution is immense. Instead, the world sees even more concentration of wealth at the top.[47] Consequently, the barriers to effective, efficient adaptation only become higher.

Moreover, even if capitalism could overcome these hindrances, the state system could well wipe away any gains in a major war, which the stresses of climate change could render more likely.[48] Another world war would push adapting to climate change off the agenda. The war would end eventually, but not before years and countless opportunities to implement adaptation had been squandered. The global average temperature would be higher, and the capacity to adapt would undoubtedly be much reduced. Even now, the emphasis on militarized security[49] means that immense resources (nearly $1.9 trillion in 2019) that could go to preventing and ameliorating damages from climate change go instead to the military. For a variety of reasons, the tendency toward high military expenditures is an element of capitalism, not only of the state system.[50] This misallocation of resources presents another barrier to successful adaptation.

Even if capitalism could, in principle, adapt effectively, the record offers no reason to believe it will. For example, mechanisms already exist for transferring funds from developed countries to developing countries to help the latter adapt. As one might expect, the funding is woefully inadequate and never likely to meet the estimated $1 trillion per year needed. Meanwhile, despite some shift toward adaptation, the focus remains on mitigation, where policy is clearly inadequate.[51] Presumably, governments could impose long-term planning, regulations, and taxation schemes to push the capitalist market system toward sufficient adaptation, but the real world indicates little hope that they will do so beyond a bit of tinkering at the margins. Public policy could aim to ensure that adaptation is equitable, so that protections against the effects of climate change benefit rich and poor alike. But a capitalism in crisis, as it will be in a world of climate catastrophe, is not likely to incline toward redistribution in favor of the poor.

Global capitalism will define the word *adaptation* as whatever ensures that capital accumulation can continue without pause; Fred Magdoff and Chris Williams assert, "To Wall Street, which sorts risks by its ability to maximize profit, 'adaptation' to climate change simply means learning how to profit from it."[52] Indeed, investors may attempt to capitalize on the risks of damage from climate change, even the risk of the collapse of civilization. As Jonathan Nitzan and Shimshon Bichler observe, "As long as their capitalization keeps rising, they will happily go under with their thumbs up."[53]

Forms of adaptation that do not support or that run against capitalization's quest for a continuous and growing stream of future income to the capitalists will be dismissed. Adaptation will be addressed primarily in ways that enhance the income stream to capital.[54] Whether such adaptation will prevent serious social consequences, even including the collapse of civilization, will be incidental to sustaining capitalization to the end.

We already see this in the implementation of adaptation today. Foreign direct assistance for adaptation falls well short of need, while fund managers hope that public funds can be used to leverage private sector investment in adaptation measures. That is, current adaptation policy maintains commitments to prevailing economic arrangements over effective responses to the growing climate crisis. Despite the net benefits of adaptation, funding fails to match the need or opportunity,[55] plainly because there is no profit in it, social benefits notwithstanding.

Support for nature-based adaptation might indicate embrace of a significantly different approach, but funding for nature-based adaptation is minimal and its efficacy is limited. The UN Environment Programme estimates financial need for funding nature-based adaptation alone will rise to $500 billion per year by mid-century, far more than is now expended or will likely ever be allocated. Even more telling is that nature-based adaptation is unlikely to succeed if global average temperature rises to high expected levels by mid-century and beyond. Nature-based adaptation depends on the success of mitigation, which is increasingly unlikely.[56] The IPCC asserts, "Above 1.5°C global warming level, some Ecosystem-based Adaptation measures will lose their effectiveness in providing benefits to people as these ecosystems will reach hard adaptation limits."[57] If some ecosystem-based adaptation will fail at the lower policy threshold, then surely most such adaptation will fail as temperatures rise toward 3.0°C by the end of this century.

To the extent capital accumulation requires economic growth, resources will deplete and the capacity of the global commons to absorb pollutants will diminish, "but capital, nature's cancer, tolerates neither limit or boundary."[58] Planetary boundaries will be transgressed.[59] On "Hothouse Earth," adaptation will require far-reaching responses to a world in a climate crisis.[60] But fundamentally, profits depend on capitalist firms not being responsible for paying for the damages they create. Nancy Fraser observes, "The effect is to incentivize owners bent on maximizing profits to commandeer nature's gifts as cheaply as possible, while also absolving them of any obligation to replenish what they take and repair what they damage."[61] Successful adaptation will almost certainly require policies and practices contrary to capital accumulation. Thus, neoliberalism's "current development pathways" must be abandoned. Daniel Uncapher and Carolyn Yvellez state, "If adaptation means simply spending a portion of the profits from the financial exploitation of our climate on ever-bigger sea walls around Manhattan, then capitalism has already lost the battle against climate change, and the 21st century promises to see cascading human rights failures of unimaginable scope and consequences."[62]

In sum, the critics say, adaptation under the rule of capital is not likely to do well going forward, arguing as follows. The inherent inequality of capitalism ensures that increasingly heavy costs will fall on those least able to bear them and with least responsibility for creating the problem. This is clearly the case today, as poor and developing countries absorb the impact of higher temperatures, and it will be increasingly the case for low-income people in the core countries as well. The fact that adaptation is implicitly defined as maintaining and enhancing the conditions for capital accumulation will become increasingly blatant. As the class basis for adaptation is revealed, resistance to bearing those costs will rise. Thus, as social and economic structures fray, more force will have to be exerted to sustain adaptation that ensures the accumulation of capital. Capitalism has failed to mitigate global warming and lacks the attributes required to adapt successfully to a crisis of its own making. Adaptation might not be so inadequate as to mean the end of civilization as such,[63] but the barriers to adaptation might well mean the end of *this* civilization. What next?[64]

Migration and Maladaptation

Before moving to a discussion of alternative social orders that might respond to the adaptation challenge better than global neoliberal capitalism guided by EMT, two significant topics must be discussed. First, this chapter challenges the notion that migration represents a potentially successful mode of adaptation to climate change.[65] Second, it investigates the meaning of maladaptation.

Migration has been proposed as an appropriate adaptation,[66] an argument that tends to see adaptation solely as an economic question. Matthew E. Kahn, for instance, identifies two important positive effects of migration to escape intolerable conditions: higher income for migrants who choose to leave a place in distress to find more suitable locales. And the quality of life for those left behind will likely rise: housing rents will decline, wages will increase, and rural people will have more land per person.[67] Both home and host countries benefit, by this account. Developing countries can alleviate population pressures in severely affected regions, and they can invest in education and training to make their citizens more competitive in the world's labor markets. If host developed countries do not restrict immigration, "then poor people in the developing world will have a much better chance to adapt to emerging threats facing their location of origin,"[68] with adaptation here understood as escaping the problem. Families in home countries can benefit if young immigrants send remittances, enabling those remaining in climate-stressed regions to purchase protections such as air conditioning and better quality food and housing. Host nations would gain from a larger pool of young workers, consequent greater capacity to fund retirement programs, and increased tax revenues. Turning everything over to markets, Kahn argues for an immigration system based on selling the right to immigrate to the United States and other developed countries, which would provide the incentives to enhance human capital and allocate immigration efficiently across destination countries. Consequently, migration, he writes, would offer a beneficial way to adapt to climate change and should be as unrestricted as possible. However, this proposal faces opposition across the political spectrum. Right-wing environmentalism focuses on ending immigration, not increasing it.[69] Meanwhile, criticizing market-based policies for environmental migration, Peter F. Cannavó writes, "Under this view, places and homes are little more than commodities—you can simply trade one for another if the price is right."[70]

A wider view of migration in response to climate change encompasses the value of place and home, as well as the impacts on communities and cultures. Cannavó's depiction of the aftermath of Hurricane Katrina that hit New Orleans, Louisiana, demonstrates the errors of commodifying place and home. Rather than costing residents only money, Katrina shattered a community, a culture, and a way of life. The refugees from Katrina, even with the meager help afforded by the government of the richest country in the world, lost networks of friends, family, and local organizations that sustained them economically and emotionally, trading them for sleeping on someone's couch while they sought employment in Houston, San Antonio, and beyond. The population of New Orleans in 2020 remained 100,000 lower than it had been in 2000, and the decline of the settled communities in the city raised fears of New Orleans losing its unique character. This is the tragic story of one city in a developed country when the stresses of climate change were considerably lower than they are today and will be in the future.

Multiply the New Orleans experience by all the impacts of climate change, include communities in developing countries, and ratchet up the level of damage every year: the loss of culture, place, and home will be immense. Migrating in search of a better job, and even having to pay the host country for the dubious privilege, would seem more a desperate attempt to survive than successful adaptation. Moreover, as the impacts of climate change become widespread, more people will seek to move to fewer viable places. The premise of migration as adaptation—that there is always somewhere to go—will weaken considerably.[71] Thus, write Kira Vinke et al., migration is far from morally neutral, and the term *adaptation* should not serve as a smokescreen for policy failures that increase injustice and suffering.[72] Thus, considering both the economic dimension and the cultural aspect, migration is a fraught solution to the climate challenge.

Beyond that, in some cases, migration as adaptation will mean the entire dissolution of the community, and this can look like a form of genocide. Consider that the definition of genocide in the Convention on the Prevention and Punishment of the Crime of Genocide includes "deliberately inflicting on the group conditions of life calculated to bring about its physical destruction in whole or in part." Worsening impacts of climate change certainly alter the conditions of life for the worse. Scattering the members of a community in response to climate risks and stresses would likely bring about the destruction of that community's collective life and identity, and it will only rarely be reconstituted in a new place. When tens of thousands fled New Orleans, they did not congregate in a new, less vulnerable place and reconstruct the cultural patterns of the city they had left. Nor would climate refugees from all parts of the world move en masse to appropriate new locales where their identity as a cohesive society is maintained. In short, the culture will be lost, not made more resilient or less vulnerable. Consequently, in some cases, migration is far from representing successful adaptation to climate change; rather, it is the failure of adaptation. While identifying a guilty party deliberately aiming to destroy a group, community, or society might be difficult, the outcome is the same: unchecked global warming met with mass migration will look like, if not be, genocide.

Is All Adaptation to Save Capitalism Maladaptation?

In the Sixth Assessment Report (AR6), the IPCC advances the goal of achieving climate-resilient development. "Climate resilient development (CRD) is a process of implementing greenhouse gas mitigation and adaptation options to support sustainable development for all," states the IPCC.[73] One important criterion for successful CRD is that measures taken to meet the climate challenge should lead to transformation. A high level of CRD encompasses system transitions and transformation, while a low level of CRD means redoubled commitment to "entrenched systems" along with unsustainable development and maladaptation.[74] Achieving CRD, says the IPCC, will require removing the soft limits to adaptation, including financial, governance, institutional, and policy barriers.

Maladaptation occurs when an adaptation measure puts a society on a path toward greater vulnerability and risk, and to outcomes that would be costly to remedy and would worsen inequality. Maladaptation can also mean adaptations adopted by one society or community that impose costs on others. Broadly, maladaptation encompasses adaptation measures that thwart progress toward CRD. Avoiding maladaptation relies on processes similar to those recommended for achieving CRD: flexible, multisector and multiactor planning that takes conscious account of the risk of maladaptation. Maladaptation often results when decisionmaking is focused on sectors or regions in isolation and fails to include all actors, especially the marginalized. If current development pathways might stand in the way of CRD, it follows that they foster maladaptation, possibly locking the world into measures that worsen conditions in the future, render remedies very costly, and impose costs inequitably. Mark Pelling suggests just such a possibility, noting that incremental adjustments could "offset immediate risks while the system itself moves ever closer to a critical threshold for collapse."[75] Consequently, any adaptation measure that preserves and prolongs the institutions, policies, and practices supporting current development pathways could be a form of maladaptation.

Analysis along these lines implies that successful adaptation cannot be limited to incremental measures to protect sectors or locales in isolation; instead, transformational change will be necessary to achieve CRD. If transformation means more than capitalism with renewable energy or putting up seawalls around Manhattan, then it must entail significant institutional change. Accordingly, even without guidance on where transformation goes, CRD implies that we must leave behind the social, political, and economic arrangements of global neoliberal capitalism or else maladaptation will result. For a given adaptation to count as successful, asserts the IPCC, it must not thwart, and should actively promote, transformation toward long-term sustainability. And although the IPCC does not say it, that means any measures taken to preserve and maintain extant institutions constitute maladaptation; rather than bring about transformation toward CRD, it would perpetuate unsustainable practices that will be unable to adapt successfully to the much hotter world to come. Put bluntly, saving capitalism is maladaptation.

Adaptation in the Web of Life

Returning to alternatives, Jason W. Moore's ambitious theoretical and empirical work takes a broad view of capitalism's remarkable capacity to adapt to constantly changing social and ecological conditions.[76] Moore suggests that capitalism must rely on four essential cheap inputs to sustain profits. In addition to cheap labor, the current configuration of capitalism requires cheap food, energy, and raw materials.[77] Traditional Marxist theory argues that value and profits result from the application of labor power to produce commodities for sale on a market. Moore claims that all four forms of cheap nature are necessary prerequisites to the production of value. Thus, "in systemic terms, Cheap Nature is produced when the interlocking agencies of capital, science, and empire—blunt categories, yes—succeed in releasing new sources of free or low-cost human and extra-human natures for capital."[78]

This means capitalism must constantly reconfigure the social-natural relationship when an existing arrangement confronts limits on the acquisition of the required low-cost inputs. Under these conditions, capital seeks new sources and types of cheap nature. To thrive, capitalism must appropriate unpaid and low-cost work and energy existing outside the system of commodity exchange. Consequently, when capitalism has confronted limits to profitability resulting from the rising cost of labor, energy, food, and natural resources, it has found new fields for appropriation of the required inputs to commodity production, whether a technological breakthrough, colonization, or mobilization of new pools of labor. It has succeeded in doing so for nearly 600 years.

The critical question now is whether global capitalism can adapt in yet another iteration to extend its existence or has, for some reason, reached the end of the line. Along with many other social, economic, political, and ecological pressures, climate change ranks as one of the most challenging barriers to capitalism's adaptation to new circumstances, and it appears that every path out of the economic crisis only exacerbates the ecological crisis. Moore writes that capitalism must constantly overcome the "exhaustion" of natural sources of profit, and "the history of capitalism has been one of recurrent frontier movements to overcome that exhaustion, through the appropriation of nature's free gifts hitherto beyond capital's reach."[79]

The question, then, is whether capitalism has now reached a juncture at which it cannot open new and low-cost energy, lands, labor, and resources for exploitation and appropriation so as to ensure cheap inputs into the production of commodities. In particular, one must ask to what extent limits are due to the impacts of climate change. Writing shortly after the financial crisis of 2008, Moore saw signs of just such a juncture. The long-run trend toward declining growth of labor productivity along with rising prices for energy and resources, evident in the aftermath of the financial meltdown, marked a "signal crisis" in which too much capital finds too few investment opportunities while production costs rise. Moore asserted that the neoliberal era of modern capitalism suffers from the lack of options for a "productivity revolution" that would restore profitability. Instead, capitalism increasingly relies on coercion for extracting value, a condition that is only likely to worsen.[80]

On this view, the definition of adaptation within capitalism would be finding paths to reducing the costs of labor, food, natural resources, and energy—the components of the four cheaps. Recent experience indicates problems in each area. Global inflation has raised prices for food and energy, therefore also raising labor costs. Climate change adds to the difficulties. IPCC reports and numerous studies project that agricultural productivity will fall as temperatures rise, at the same time that global population and the demand for food increase. Consequently, food is likely to become relatively more expensive. Renewable energy has become competitive with fossil fuels, which could allow continuation of the low-cost energy regime that has driven global capitalism without necessarily causing a massive increase in GHG emissions. However, deploying renewable energy carries significant costs, including prices of minerals needed to produce batteries, meaning another of the four cheaps could see rising costs. Moreover, it is unlikely that the global energy system will abandon fossil fuels. Instead, as noted, consumption of fossil fuels will remain too high to meet climate goals even while the share provided by renewables increases; this is perhaps the most likely path for holding energy costs down for as long as possible, but at the cost of higher concentrations of GHG in the atmosphere. As temperatures increase, labor productivity falls, a particular concern for agriculture, meaning in turn higher food prices.

In sum, the factors Moore identifies as critical to the health of global capitalism are all under added pressure from climate change, and no path to restore low costs for food, labor, natural resources, and energy is in view. Doing so would be the prerequisite for implementing many forms of incremental adaptation, from early warning systems to seawalls and more. But no clear way forward exists, and that means all manner of specific adaptation measures are unlikely to be implemented. If this depiction of the current situation and future prospects is broadly correct, then capitalism's remarkable adaptability will have reached its limit. Capitalism will pose hard limits on adaptation. Moore asserts, "The ecological limit of capitalism, it appears, is capital itself."[81]

Moore's analysis points to a probable scenario in which it is difficult to impossible to reduce the cost of natural resources, energy, or food. That leaves one cheap that could allow profits to grow: labor. Labor power must be reproduced to keep the assembly lines running, but what is required or expected to enable that reproduction can vary widely. The rising cost of food and energy can put upward pressure on wages, but resistance to wage increases can mean instead that workers reduce expenditures on other necessities and luxuries such as quality housing and entertainment. At the limit, workers can be reduced to the cheapest kinds of housing (barracks, for instance) and simply forgo any expenditure that is not for vital necessities. Thus, the bulk of the returns from commodity sales would go to the owners of the factories, mines, and the like. However, this might only bring about a crisis of overproduction as goods on the market go begging for buyers. What is more, workers are unlikely to accept deteriorating conditions and will push back with individual choices about where to work and collective efforts such as unionization and strikes. Thus, forcing wages down can only go so far before workers resist with ever stronger measures. With no new fields available for cheap energy, food, or resources, and the limits of labor exploitation reached, capitalism could no longer adapt so as to perpetuate itself.

Before that moment is reached, the class nature of the climate crisis will likely be evident. When capitalism's capacity to appropriate cheap energy, food, and natural resources reaches its limit, society simplifies to a stark confrontation between the owners of capital and the workers. In 1848, Marx and Engels wrote that capitalist society had "simplified the class antagonisms," but nearly two centuries later, that direct confrontation had not yet resulted in the end of capitalism. Moore's analysis indicates one reason for the seemingly endless deferral is that capital has adapted in ways that allow it to avoid the crisis, turning to lower costs for energy, food, and natural resources to sustain profits without complete immiseration of the working class. However, if, as Moore suggests, those roads are now, or soon will be, blocked, then all that remains is the "splitting up into two great hostile camps, into two great classes facing each other – bourgeoisie and proletariat." This confrontation will result "either in a revolutionary reconstitution of society at large, or in the common ruin of the contending classes."[82] Adaptation from this perspective, then, is about seeking ways to avoid the simplification of society such that the only alternatives are complete transformation of society or complete ruination of the contending classes. To the extent climate change presents barriers to cheapening inputs to production, by the same token it hastens the hour of reckoning when capitalism must either give way to another social system or be destroyed. For civilization to survive and thrive, global society must undergo profound transformational change. Where might such a transformation take us?

Notes

1. Joe Lo, "Kerry Rejects 'Climate Reparations' but Praises Loss and Damage Fund," *Climate Home News,* July 14, 2023.

2. Johanna Wolf, Karen O'Brien, and Linda Sygna, "The Future Is Now," in *A Changing Environment for Human Security: Transformative Approaches to Research,*

Policy and Action, edited by Linda Sygna, Karen O'Brien, and Johanna Wolf (London: Routledge, 2013), p. 38.

3. Kirstin Dow, Frans Berkhout, Benjamin L. Preston, Richard J. T. Klein, Guy Midgley, and M. Rebecca Shaw, "Limits to Adaptation," *Nature Climate Change* 3, no. 4 (2013): 305–307.

4. Bentley B. Allan and Jonas O. Meckling, "Creative Learning and Policy Ideas: The Global Rise of Green Growth," *Perspectives on Politics* 21, no. 2 (2022): 443–461.

5. Dana R. Fisher and William R. Freudenburg, "Ecological Modernization and Its Critics: Assessing the Past and Looking Toward the Future," *Society and Natural Resources* 14, no. 8 (2001): 701–709.

6. John Bellamy Foster, "The Long Ecological Revolution," *Monthly Review* 69, no. 6 (2017), p. 3.

7. Richard York, Eugene A. Rosa, and Thomas Dietz, "Ecological Modernization Theory: Theoretical and Empirical Challenges," in *The International Handbook of Environmental Sociology,* 2nd ed., edited by Michael R. Redclift and Graham Woodgate (Cheltenham: Edward Elgar, 2010), pp. 77–90.

8. John Bellamy Foster, "Making War on the Planet: Geoengineering and Capitalism's Creative Destruction of the Earth," *Monthly Review* 70, no. 4 (2018): 1–10.

9. Matthew E. Kahn, *Adapting to Climate Change: Markets and the Management of an Uncertain Future* (New Haven: Yale University Press, 2021), p. 75.

10. Matthew E. Kahn, *Adapting to Climate Change: Markets and the Management of an Uncertain Future* (New Haven: Yale University Press, 2021).

11. David Pepper, "Sustainable Development and Ecological Modernization: A Radical Homocentric Perspective," *Sustainable Development* 6, no. 1 (1998): 1–7.

12. Richard York, Eugene A. Rosa, and Thomas Dietz, "Ecological Modernization Theory: Theoretical and Empirical Challenges," in *The International Handbook of Environmental Sociology,* 2nd ed., edited by Michael R. Redclift and Graham Woodgate (Cheltenham: Edward Elgar, 2010), p. 79.

13. Emmanuel Macron, Mia Amor Mottley, Luiz Inacio Lula da Silva, Ursula von der Leyen, Charles Michel, Olaf Scholz, and Fumio Kishida et al., "A Green Transition that Leaves No One Behind," *Project Syndicate,* June 20, 2023, https://www.project-syndicate.org.

14. Bill Blackwater, "The Denialism of Progressive Environmentalists," *Monthly Review* 64, no. 2 (2012): 10–21; John Bellamy Foster, "The Planetary Rift and the New Human Exemptionalism: A Political-Economic Critique of Ecological Modernization Theory," *Organization and Environment* 25, no. 3 (2012): 211–237.

15. Jeff Goodell, "Human Adaptation to Heat Can't Keep Up with Human-Caused Climate Change," *Time,* July 6, 2023.

16. Carlo Fannelli, "Climate Change: 'The Greatest Challenge of Our Time,'" *Alternate Routes* 14 (2014), p. 19.

17. Laurie Parsons, *Carbon Colonialism: How Rich Countries Export Climate Breakdown* (Manchester: Manchester University Press, 2023).

18. John Bellamy Foster, "The Long Ecological Revolution," *Monthly Review* 69, no. 6 (2017), p. 13.

19. John Barry, "Green Republicanism and a 'Just Transition' from the Tyranny of Economic Growth," *Critical Review of International Social and Political Philosophy* 24, no. 5 (2019): 725–742.

20. Matthew Paterson, "Climate Change and International Political Economy: Between Collapse and Transformation," *Review of International Political Economy* 28, no. 2 (2021), p. 400.

21. Donella Meadows, Dennis Meadows, Jørgen Randers, and William Behrens III, *The Limits to Growth: A Report for the Club of Rome's Project on the Predicament of Mankind* (New York: Signet, 1972).

22. Ted Nordhaus, Vijaya Ramachandran, and Patrick Brown, "The Obvious Climate Strategy Nobody Will Talk About: Economic Development Is the Only Proven Path to Climate Resilience," *Foreign Policy,* November 6, 2022.

23. Paul R. Ehrlich and Dennis C. Pirages, "Political Science in a New Era," *World Futures Review* 4, no. 3 (2012), p. 18.

24. Paul R. Ehrlich and Anne H. Ehrlich, "Can a Collapse of Global Civilization Be Avoided?" *Proceedings of the Royal Society B* 280, no. 20122845 (20131-9).

25. Nordhaus, Ted, Vijaya Ramachandran, and Patrick Brown, "The Obvious Climate Strategy Nobody Will Talk About: Economic Development Is the Only Proven Path to Climate Resilience," *Foreign Policy,* November 6, 2022.

26. Laurie Parsons, *Carbon Colonialism: How Rich Countries Export Climate Breakdown* (Manchester: Manchester University Press, 2023), p. 4, emphasis in original.

27. Mark Harvey, "Climate Emergency: How the Inequality Crisis Is Dynamically Linked to the Sociogenesis of Climate Change," *Global Social Challenges Journal* 1, no. 1 (2022), p. 133. See also Norichika Kanie, Michele M. Betsill, Ruben Zondervan, Frank Biermann, and Oran R. Young, "A Charter Moment: Restructuring Governance for Sustainability," *Public Administration and Development* 32, no. 3 (2012): 292–304.

28. John Bellamy Foster, "Capitalism and the Curse of Energy Efficiency: The Return of the Jevons Paradox," *Monthly Review* 62, no. 6 (2010): 1–12.

29. Jason Hickel and Giorgos Kallis, "Is Green Growth Possible?" *New Political Economy* 25, no. 4 (2020): 469–486.

30. Matthew E. Kahn, *Adapting to Climate Change: Markets and the Management of an Uncertain Future* (New Haven: Yale University Press, 2021).

31. David Schweickart, "Is Sustainable Capitalism Possible?" *Procedia Social and Behavioral Sciences* 2, no. 5 (2010): 6739–6752.

32. Michael J. Albert, "Capitalism and Earth System Governance: An Ecological Marxist Approach," *Global Environmental Governance* 20, no. 2 (2020): 37–56; Nancy Fraser, "Behind Marx's Hidden Abode," *New Left Review* 86 (2014): 55–72; James O'Connor, "A Prolegomenon to an Ecological Marxism: Thoughts on the Materialist Conception of History," *Capitalism Nature Socialism* 10, no. 2 (1999): 77–106; and many others.

33. Joel Kovel, *The Enemy of Nature: The End of Capitalism or the End of the World?* (London: Zed Books, 2007), p. 82.

34. See, for instance, Achim Brunnengräber, "The Political Economy of the Kyoto Protocol," in *Socialist Register 2007: Coming to Terms with Nature,* edited by Leo Panitch and Colin Leys (London: Merlin Press, 2006), pp. 213–230; Fred Magdoff, and John Bellamy Foster, "What Every Environmentalist Needs to Know About Capitalism," *Monthly Review* 61, no. 10 (2010): 1–30; John Barkdull and Paul G. Harris, "Climate-induced Conflict or Hospice Earth: The Increasing Importance of Eco-Socialism," *Global Change, Peace and Security* 27, no. 2 (2015): 237–243; Paul Burkett, "An Eco-Revolutionary Tipping Point?" *Monthly Review* 69, no. 1 (2017): 1–19.

35. Chris Williams, *Ecology and Socialism* (Chicago: Haymarket Books, 2010), p. 229.

36. Nancy Fraser, "Climates of Capital: For a Trans-Environmental Eco-Socialism," *New Left Review* 127 (2021), p. 99.

37. UNEP (UN Environment Programme), *Adaptation Gap Report 2021: The Gathering Storm—Adapting to Climate Change in a Post-Pandemic World—Executive Summary* (Nairobi: 2021).

38. Mike Davis, "Who Will Build the Ark?" *New Left Review* 61 (2010), pp. 29–30.

39. Paul G. Harris argues that the climate policy orientation should shift from international politics among states—"the cancer of Westphalia"—to recognition that affluent people in developed and developing countries are imposing costs on less wealthy people in all countries, not only those in poor countries. See Paul G. Harris, *What's Wrong with Climate Politics and How to Fix It* (Cambridge, England: Polity, 2013), pp. 119–120.

40. Philip Antwi-Agyei, Andrew J. Dougill, Lindsay C. Stringer, and Samuel Nii Ardey Codjoe, "Adaptation Opportunities and Maladaptive Outcomes in Climate Vulnerability Hotspots of Northern Ghana," *Climate Risk Management* 7, no. 4 (2018): 83–93.

41. Chris Williams, "How Will We Get to an Ecological Civilization?" *Climate and Capitalism,* November 2, 2015.

42. John Michael Greer, *Dark Age America: Climate Change, Cultural Collapse, and the Hard Future Ahead* (Gabriola Island, British Columbia, Canada: New Society, 2016).

43. Shyam Saran provides a typical statement of this view: "Therefore, what we need is a mechanism through which we can build up the coping capacities of developing countries, particularly those that are most vulnerable." Shyam Saran, "Global Governance and Climate Change," *Global Governance* 15, no. 4 (2009), p. 458.

44. Mark Lynas, *Our Final Warning: Six Degrees of Climate Emergency* (London: 4th Estate, 2020), p. 173.

45. Stephen Zehr, "The Sociology of Global Climate Change," *WIREs Climate Change* 6, no. 2 (2015), p. 142.

46. Iris Borowy, "Sustainability and Redistribution," in *What Next for Sustainable Development: Our Common Future at Thirty,* edited by James Meadowcroft, David Banister, Erling Holden, Oluf Langhelle, Kristin Linnerud and Geoffrey Gilpin (Cheltenham: Edward Elgar, 2019), p. 125.

47. Inequality.org, "Global Inequality," https://inequality.org, n.d.

48. John Barkdull and Paul G. Harris, "Climate-Induced Conflict or Hospice Earth: The Increasing Importance of Eco-Socialism," *Global Change, Peace and Security* 27, no. 2 (2015): 237–243.

49. On the challenge of militarism, see David Roche, *The United Nations in the 21st Century* (Toronto, Ontario, Canada: James Lorimer, 2015), pp. 23–32.

50. Paul A. Baran and Paul M. Sweezy, *Monopoly Capital: An Essay on the American Economic and Social Order* (New York: Monthly Review Press, 1966).

51. IPCC (Intergovernmental Panel on Climate Change), *Climate Change 2014: Impacts, Adaptation and Vulnerability. Contribution of Working Group II to the Fifth Assessment Report of the Intergovernmental Panel on Climate Change* (Cambridge: Cambridge University Press, 2014), p. 844.

52. Fred Magdoff and Chris Williams, *Creating an Ecological Society: Toward a Revolutionary Transformation* (New York: Monthly Review Press, 2017), p. 148.

53. Jonathan Nitzan and Shimshon Bichler, *Capital as Power,* RIPE Series in Global Political Economy (New York: Routledge, 2009), p. 619.

54. The trend in mainstream conservation is indicative, as major corporations, the big nongovernmental organizations, and intergovernmental institutions enthusiastically "mainstream" "neoliberal conservation" and its practice of pricing nature

so as to save it. See Bram Büscher and Robert Fletcher, *The Conservation Revolution: Radical Ideas for Saving Nature Beyond the Anthropocene,* Kindle ed. (London: Verso, 2020), loc. 304–377.

55. UNEP, *Adaptation Gap Report 2020* (Nairobi: UNEP, 2021): "The Global Commission on Adaptation estimated that a US$1.8 trillion investment in the areas of early warning systems, climate-resilient infrastructure, improved dryland agriculture, global mangrove protection and resilient water resources could generate US$7.1 trillions of avoided costs and non-monetary social and environmental benefits," p. xiv. Yet "there is a real risk that adaptation costs will increase faster than adaptation-oriented finance.," p. xviii.

56. John Barkdull and Paul G. Harris, "Emerging Responses to Global Climate Change: Ecosystem-Based Adaptation," *Global Change, Peace and Security* 31, no. 2 (2019): 19–37.

57. IPCC, "Summary for Policymakers," in *Climate Change 2022: Impacts, Adaptation and Vulnerability. Contribution of Working Group II to the Sixth Assessment Report of the Intergovernmental Panel on Climate Change* (Cambridge: Cambridge University Press, 2022), p. 26.

58. Joel Kovel, *The Enemy of Nature: The End of Capitalism or the End of the World?* (London: Zed Books, 2007), p. 170.

59. Johan Rockström, Will Steffen, Kevin Noone, Asa Persson, F. Stuart Chapin III, Eric Lambin, and Timothy M. Lenton et al., "Planetary Boundaries: Exploring the Safe Operating Space for Humanity," *Ecology and Society* 14, no. 2 (2009), p. 32.

60. Will Steffen, Johan Rockstrom, Katherine Richardson, Timothy M. Lenton, Carl Folke, Diana Liverman, and Colin P. Summerhayes et al., "Trajectories of the Earth System in the Anthropocene," *Proceedings of the National Academy of Sciences* 115, no. 33 (2018): 8252–8259, DOI 10.1073/pnas.1810141115.

61. Nancy Fraser, "Climates of Capital: For a Trans-Environmental Eco-Socialism," *New Left Review* 127 (2021), p. 100.

62. Daniel Uncapher and Carolyn Yvellez, "Climate Adaptation Needs to Put Human Rights Above Property Values," *Truthout,* July 14, 2019.

63. See, for instance, Toby Ord, *The Precipice: Existential Risk and the Future of Humanity* (New York: Hachette Books, 2020), pp. 102–113.

64. John Bellamy Foster, "Capitalism Has Failed—What Next?" *Monthly Review* 70, no. 9 (2019): 1–24.

65. Susan Martin, "Climate Change, Migration, and Governance," *Global Governance* 16, no. 3 (2010): 397–414; Douglas K. Bardsley and Graeme J. Hugo, "Migration and Climate Change: Examining Thresholds of Change to Guide Effective Adaptation Decision-Making," *Population and Environment* 32, nos. 2–/3 (2010): 238–262.

66. Kira Vinke, Jonas Bergmann, Julia Blocher, Himani Upadhyay, and Roman Hoffmann, "Migration as Adaptation?" *Migration Studies* 8, no. 4 (2020): 626–634; Chris Jacobson, Stacy Crevello, Chanthan Chea, and Ben Jarihani, "When Is Migration a Maladaptive Response to Climate Change?" *Regional Environmental Change* 19, no. 10 (2019): 101–112.

67. Matthew E. Kahn, *Adapting to Climate Change: Markets and the Management of an Uncertain Future* (New Haven: Yale University Press, 2021), pp. 224–229.

68. Matthew E. Kahn, *Adapting to Climate Change: Markets and the Management of an Uncertain Future* (New Haven: Yale University Press, 2021), p. 224.

69. Oliver Milman, "Climate Denial Is Waning on the Right. What's Replacing It Might Be Just as Scary," *The Guardian,* November 21, 2021.

70. Peter F. Cannavó, "In the Wake of Katrina: Climate Change and the Coming Crisis of Displacement," in *Political Theory and Global Climate Change,* edited by Steve Vanderheiden (Cambridge: MIT Press, 2008), p. 178.

71. Jonathan Mingle, "Forget About Hiding from Climate Chaos in America," *New York Times,* July 13, 2023; Sarah Kaplan, "Floods, Fires and Deadly Heat Are the Alarm Bells of a Planet on the Brink," *Washington Post,* July 12, 2023.

72. Kira Vinke, Jonas Bergmann, Julia Blocher, Himani Upadhyay, and Roman Hoffmann, "Migration as Adaptation?" *Migration Studies* 8, no. 4 (2020), p. 631.

73. IPCC, "Climate Resilient Development Pathways," in *Climate Change 2022: Impacts, Adaptation and Vulnerability. Contribution of Working Group II to the Sixth Assessment Report of the Intergovernmental Panel on Climate Change* (Cambridge: Cambridge University Press, 2022), p. 2657.

74. IPCC, "Summary for Policymakers," in *Climate Change 2022: Impacts, Adaptation and Vulnerability. Contribution of Working Group II to the Sixth Assessment Report of the Intergovernmental Panel on Climate Change* (Cambridge: Cambridge University Press, 2022), p. 30.

75. Mark Pelling, "Resilience and Transformation," in *Climate Change and the Crisis of Capitalism: A Chance to Reclaim Self, Society and Nature,* edited by Mark Pelling, Daniel Manuel-Navarette, and Michael Redclift (London: Taylor and Francis, 2011), p. 53.

76. Jason W. Moore, "Ecology, Capital, and the Nature of Our Times: Accumulation & Crisis in the Capitalist World-Ecology," *Journal of World-Systems Research* 17, no. 1 (2011): 107–146.

77. Jason W. Moore, *Capitalism in the Web of Life: Ecology and the Accumulation of Capital,* Kindle version (London: Verso, 2015).

78. Jason W. Moore, *Capitalism in the Web of Life: Ecology and the Accumulation of Capital,* Kindle version (London: Verso, 2015), loc. 1259.

79. Jason W. Moore, "Ecology, Capital, and the Nature of Our Times: Accumulation & Crisis in the Capitalist World-Ecology," *Journal of World-Systems Research* 17, no. 1 (2011): 107–146.

80. Jason W. Moore, "Ecology, Capital, and the Nature of Our Times: Accumulation & Crisis in the Capitalist World-Ecology," *Journal of World-Systems Research* 17, no. 1 (2011), p. 137.

81. Jason W. Moore, "Ecology, Capital, and the Nature of Our Times: Accumulation & Crisis in the Capitalist World-Ecology," *Journal of World-Systems Research* 17, no. 1 (2011), p. 110.

82. Karl Marx and Friedrich Engels, "The Communist Manifesto (1848)," in *The Portable Marx,* edited by Eugene Kamenka (New York: Penguin Books, 1983), p. 204.

6

Defending Civilization as We Know It

VISIONS OF A RADICALLY TRANSFORMED SOCIETY DIVIDE into two broad groups: those seeking ways to perpetuate civilization, and those saying this civilization has run its course. This chapter reviews the first group, and Chapter 7 addresses the second group.

The Steady State

"Much of the modern environmental debate," say Damien F. White, Alan P. Rudy. and Brian J. Gareau, "has been defined by where one stands on the question of *limits*."[1] The notion of the steady state addresses this issue by asserting that limits do exist, and a sustainable society must maintain a balance between the uses people make of nature and nature's capacity to replenish itself.[2] While development can be sustainable, Herman Daly asserts, growth cannot. The human economy exists within a finite ecosystem of fixed size, so the economy cannot perpetually grow, and "sustainable growth" is oxymoronic. We require instead sustainable development, which is development without endless material growth.[3]

Economic activity should respect two limits: the rate of use of natural resources should not exceed sustainable regeneration, and waste emission rates should not exceed the planet's capacity to assimilate those wastes. Regarding nonrenewable resources, such as oil, sustainability requires that as nonrenewables are used, there must be a compensating investment in renewable substitutes such as solar power. Technological change should be guided by the rule of enhancing efficiency rather than growth; for example, instead of building a new power plant, install more efficient electrical devices. Regarding poverty, the solution cannot be growing the world economy by five- or tenfold; it will require population control and significant redistribution of wealth. The steady state, calibrated to the carrying capacity of regional and global ecologies, will allow sustainable development, but rejects "sustainable growth."

Advocates of the steady state are clear about the need for a road map to the future: "If humanity is to achieve a sustainable and desirable future, we must create a shared vision detailing what we as a society want to sustain and incorporating the central shared values that express our hopes for the future."[4] What does that shared vision entail? Economic production in the steady state society would focus on quality, not quantity, and evaluate economic activity in regard to the ends—sustainable human well-being—rather than the means, so that society can develop without physically growing. Urban areas will have been redesigned so most needs can be met within a twenty-minute walk, public transportation is readily available, green areas are abundant, and buildings are built to last. Renewable energy would power society, and industrial production would be local and nonpolluting.

Companies would mostly be worker owned or public enterprises. Business competition would still exist, as would even some large corporations, but enterprises would be more socially responsible and include broader representation on boards of directors. The profit motive would not dominate business decisions. Enhancing human capital would be understood as fostering the full development of the individual's health, well-being, talents, and sense of community, with both education and work changing to reflect society's orientation toward true sustainability. Social capital would bind the postgrowth community, replacing individualism and acquisitiveness with strong moral bonds of honesty, fair dealing, care for the less fortunate, and shared cultural expectations and practices to which people adhere voluntarily. In sum, the steady state requires a new outlook to support extensive social reforms. Achieving the steady state will mandate stable populations, equitable sharing of resources, income equality, appropriate work, investment in the commons, financial reform, tax incentives toward sustainability, technology for human well-being, democracy, and better measures of progress. "In other words, a complete makeover."[5]

Without question, this plan for the steady state would involve significant change in social, economic, and political institutions, in addition to requiring leaders and citizens to adopt a very different view of society and nature. However, the suggestion to "reform" the financial system and the tax system indicates that Alperovitz et al. expect those institutional arrangements to be there. Much of the weight of the proposal rests on the adoption of appropriate values and worldviews so existing institutions can be directed toward sustainability.

Retaining growth-oriented institutions while striving for a no-growth society could represent a contradiction at the heart of the steady state proposal. If many critics of capitalism are correct, some elements of society would constantly be pressing for their own expansion, subverting the overall goal of no expansion. Growth would not be an option for resolving a crisis of underproduction and overconsumption, which would drive the system to crisis. Necessarily, firms will constantly press against the system's boundaries; those that do not will be eliminated by competitors. To the extent the steady state plan for the future relies on today's corporate structure and the various agencies and international organizations supporting their activities, it will face an unmanageable tension at the heart of the system.

That said, if the proposal to democratize the economy by implementing local workers control and ownership were fully implemented, then the relations of production and thus the rest of society would have been profoundly changed. In that case, however, there would be little need to discuss reform of financial and tax systems; those systems would be radically transformed. By the same token, technological change would occur on widely different principles than currently, likely serving human well-being as the logical consequence of those principles. Meanwhile, the proposals to expand and deepen democracy would represent a significant transformation of the political system. The combination of the steady state's striving for equity, deep democracy, worker ownership of enterprises, and production for human needs, taken to its ultimate outcome, might be called *ecosocialism*. However, advocates of the steady state tend to avoid that label.

What would the steady state mean for adaptation to climate change? The answer to that question would depend in part on how quickly the world achieved the steady state, for that would determine how much higher temperatures go. To have a good chance to remain below 2.0°C or, better, 1.5°C, would require developed countries to eliminate nearly all greenhouse gases (GHG) emissions almost overnight. Developing countries would have to commit to energy futures based on renewables, with some time allowed for poverty alleviation while they make the transition. As we have seen, the more likely path is toward about 3.0°C temperature rise by the end of this century. Current experience with 1.0°C rise suggests that 3.0°C will be catastrophic. The plan for the steady state tackles this problem directly, insisting that the steady state requires net zero GHG emissions, and the steady state plan is no more plagued by the practical improbability of getting there than any other proposal for changing society. But say there is a brisk transition to the steady state, contributing to holding the global average temperature rise to 2.5°C. A steady state society would still have to adapt to climate impacts at that level. Accordingly, the steady state, as well as other proposals outlined below, will face severe climate impacts, including rising seas, dangerous heat, flooding in some areas and drought in others, mass species loss and extinction, and a possible "major structural food deficit."[6] At that temperature, the world might have reached tipping points that ensure global warming would continue for decades beyond the 2100 policy horizon.

Robert Costanza et al. note that adaptation will have to be part of the effort to turn away from consumerism: "Shifting the traditional focus of investment towards renewable energy, public goods, green (resource-efficient) technology, climate adaptation and mitigation, etc."[7] This is the only direct reference to climate adaptation in a document describing a steady state society, and it appears to underplay the magnitude of the issue, relegating it to one of the items to be addressed in a reorientation of investment priorities. Similarly, the steady state would require better measures of human well-being than gross domestic product (GDP), tax reform, enhanced resilience, and expansion of the commons relative to the private sector; again, not sweeping transformation.[8] Still, taking a broader view of the matter suggests that the design for a steady state society is itself the transformational change required to adapt equitably and effectively to climate change. Presumably, a world that is more democratic, features worker-owned enterprises, allows social and ecological concerns to determine investment decisions rather than only the profit motive, and is consciously striving for ecological balance would be capable of making the right decisions on how to adapt to now-inevitable global warming.

Degrowth

Some analysts argue that the steady state is not sufficient. Instead, they say, the global human population has already overshot the planet's carrying capacity, and to achieve sustainability will require reducing the size of the economy and the population to bring human activity in line with ecological limits. That is, beyond achieving the steady state, society will have to shrink to avoid catastrophe. Degrowth, assert advocates, does not mean simply inducing recessions or depressions. Instead, it is an intentional process to plan comprehensive institutional changes geared toward reducing production and consumption while increasing human welfare and equity.[9]

The degrowth perspective forwards a critique of development, arguing that, whether carried out by market-friendly governments, socialists, or humanitarian nongovernmental organizations (NGOs), development "has been an overwhelming failure,"[10] often enough making the people subjected to it worse off, regardless of aggregate statistics showing a growing GDP. Likewise, degrowth rejects the notion of green growth, citing evidence that decoupling has not happened in the economy as a whole and cannot happen fast enough to avert dangerous climate change and other environmental problems.[11] Degrowth is focused on reducing material throughputs to a sustainable level, ignoring such aggregate measures of economic activity as GDP. In practice, degrowth would likely slow the rate of growth of GDP, and it might reduce GDP. But because degrowth is concerned with other aspects of human well-being, this will not constitute a crisis as it does in a growth-oriented model; in pursuit of the greater goal of reducing throughputs, degrowth suggests ways of managing the difficulties that will arise from slow growth or declining GDP.

The term *growth,* writes Jason Hickel, is generally considered a good thing, but it obscures that what is really happening is elite accumulation of wealth, commodification of the commons, and appropriation of labor and resources.[12] By contrast, degrowth, as a critique of capitalism, calls for disaccumulation, decommodification, and decolonization. Degrowth theorists acknowledge the needs that are remaining in poor countries, so their recommendations for reduced and reoriented production are aimed at the developed countries. Indeed, says Hickel, the environmental damages occurring in the developing world are largely attributable to those areas' economic ties with the developed countries, and they would be better off economically and environmentally if developed countries relied on them less.[13] Unfortunately, this option has been blocked. Instead, the turn toward neoliberalism by the wealthy countries in the 1980s was intended to open up more areas to serve the needs of capital accumulation, putting an end to developing countries' efforts to forge independent development paths. In line with Moore's analysis of the four cheaps,[14] neoliberalism represented yet another round of capitalism's transformations to sustain profitability and thereby the global structure of class relations. However, neoliberalism is now exhausted and, as noted, no obvious avenue for restoration of the conditions for further growth is available. Extracting the last bit out of the remaining resources, labor, and land of the Global South means that growth in the developed world will result in even further devastation of the developing world. Rather than save a faltering system at such cost, degrowth, concentrated in the Global North, would put rich countries on the path to sustainability while removing a large part of the environmental burden carried by developing countries.

Critics of degrowth include ecosocialists who have argued that degrowth refuses to grapple with the capitalist world order's entire dependence on the growth imperative.[15] Failure to do so leaves degrowth with no realistic way to implement its principles. The institutions and supporting social practices would remain, somehow expected to adopt entirely new purposes in line with degrowth recommendations. But as we have seen, many observers, including ecosocialists, believe growth is essential to capitalism and there is no way to reconcile capitalism and degrowth. Thus, assuming degrowth is the right path, it remains to outline the institutional transformations that would enable degrowth principles to be put into practice.

Degrowth advocates have responded to this criticism. They assert that they expect something other than a shrinking capitalism. A global order based on degrowth would have to dismantle much of the apparatus of capitalism, including irrational long-distance trade, growth-obsessed corporations, the global financial system, and the many international organizations and agencies that facilitate the accumulation of capital. In addition, degrowth will require a far more democratic decisionmaking process, one with real content rather than the elite-managed form of democracy seen in the world today. Most degrowth advocates are sympathetic to socialism, but socialism without growth. The old notion that socialism would mean new managers would take over the industrial capacities of society, run them to benefit workers, and grow production faster than the profit-driven corporate management could manage must be abandoned.

Furthermore, in response to those who say reversing growth would impoverish working people, advocates respond that degrowth is not the same as a permanent recession in the global economy. Instead, they say, degrowth is a planned program to achieve sustainability, reduce inequality, and enhance human well-being. Unlike unplanned economic slumps, degrowth would target reductions in wasteful and ecologically harmful activities while encouraging growth in health care, education, and social solidarity—all of which are harmed in recessions while corporations grow and profit. Degrowth would seek to ensure employment for all, contrary to a capitalist recession in which unemployment soars. Presumably, degrowth would reject the use of higher interest rates and constricting the money supply to induce unemployment so as to tamp down inflation, contrary to what the Federal Reserve has done in the United States since the Covid-19 pandemic. Degrowth would aim especially at reducing inequality, to alleviate poverty while abandoning the GDP growth obsession, and it would expand public services to ensure a decent life for all. Most pertinent for this study, "Degrowth is part of a plan to achieve a rapid transition to renewable energy, restore soils and biodiversity, and reverse ecological breakdown."[16] In short, degrowth is not simply the opposite of growth, shrinking GDP instead of increasing it. It is instead a deliberate, planned process of reorienting the economy and society to stay within planetary boundaries while providing more well-being.

What would degrowth mean for adaptation to climate change? Reducing the material throughput would directly address mitigation, as a degrowth society would forgo the use of fossil fuels and transition to renewables as quickly as possible. Indeed, it is expected that new jobs in the renewable energy sector would partly answer the question of unemployment in a shrinking economy. The more successful this effort, the less temperature increases and thus less adaptation would be required. Further, a political and economic system not tied to endless growth, one that incorporates other values than profits in its policy and economic choices, could opt for equitable, sustainable adaptation measures, and it would be more likely than current arrangements to avoid maladaptation. Similarly to the steady state, transformational change in the entire basis for society would be the foundation for adaptation to the now-unavoidable effects of climate change, unavoidable regardless of how effective a degrowth society's mitigation efforts might be. A properly organized society, geared toward reducing society's ecological impact while ensuring well-being for all, would allocate resources without regard for profits, to ensure the survival and health of communities threatened by rising temperatures. The precise mechanism for transformed decisionmaking is not described in detail, but it would be genuinely democratic.

Degrowth remains subject to criticism for failing to depict the institutional and organizational basis of the new society.[17] Critiquing productivist socialism, Serge Latouche asserts, "In short, we must destroy industrial society."[18] To be sure, capitalism, which requires accumulation, depends on the exploitation of labor and "the endless destruction of nature," and the degrowth society must therefore reverse accumulation and exploitation. It must also "challenge the logic of capitalism." However, Latouche continues, the anticapitalist left must also be challenged, for it too often embraces the view that the growth of employment, production, and consumption are good, while the growth of accumulated capital is the source of all evil. A socialist barrel of oil, he notes, is just as polluting as a capitalist barrel of oil, so the reorganization of society must go beyond changing the identities of the managers of growing industrial production. But with both of the most compelling models of society rejected, what is the alternative? Degrowth provides principles, but not institutional specifics. "But degrowth in itself," Michael Lowy writes, "is not an alternative economic and social perspective: it does not define what kind of society will replace the present system."[19] Ecosocialism fills that gap.

Ecosocialism

Ecosocialists begin with the critique of capitalism already reviewed here: it is heedlessly addicted to growth regardless of ecological costs, and capitalism has no way out, so it is hurtling toward ecological limits and will bring about a catastrophe. Whereas a rational response to planetary boundaries would reorient from consumerism to human needs, from corporate dictatorship to deep democracy, and from private ownership of the means of production to collective ownership and management, capitalism is inherently blocked from such a rational ecological response. Historically, capitalism has expanded because fossil fuels offer cheap energy—cheap, but nonrenewable and ecologically damaging. Lowy writes, "Capitalism cannot exist without growth, expansion, accumulation of capital, commodities, and profits, and this growth cannot go on without an extended use of fossil fuels."[20] Thus, today's predominant economic and political arrangements are unlikely to implement mitigation quickly enough to avert catastrophic climate change, putting the onus on capitalism's capacity to adapt, which is also doubtful, as outlined above.

The important question to be analyzed in this section is what a postcapitalist ecosocialist world order would look like. Ecosocialists assert that the capitalist mode of production is incapable of having a sustainable, harmonious relationship with nature, so what organization of society do they recommend? Since Karl Marx, socialists have focused the analysis of capitalism on the class relationship embodied in social practices; namely, the opposition between the owners of society's productive capacities (the capitalists or bourgeoisie) and the people who sell their labor power for a wage (the working class or proletariat). Those who own the factories, mines, offices, shopping centers, airlines, and the rest, and who manage the process of capital accumulation, have the power to exclude individual workers from a job and thus to the means of subsistence. Consequently, the owners largely dictate the terms, with workers, citizens, and communities having little say in the matter. This disempowers the mass of society, the people who would have the greatest interest in a healthy environment, and turns the productive capacities of society toward enriching the owners regardless of ecological cost. Consequently, an ecosocialist order would do away with this class relationship by eliminating the owning class, seizing control over the society's productive capacities, and putting the means of production under the control of the associated producers and citizens.

No longer driven by the imperatives of growth and capital accumulation, the new economic managers would enact policies and practices to bring human society into harmony with the natural world and achieve true sustainability.[21] Massive redistribution of wealth would occur, obviating the need for endless economic growth to alleviate poverty. Production in wasteful and ecologically harmful industries would cease; this includes the so-called defense industry, luxury goods for the super rich, and the vast advertising industry with its goal of creating consumer demand for frivolous products. Meeting the environmental challenge, including mitigating and adapting to climate change, would require transforming society's energy system, transportation, housing, urban planning, agriculture, and industry. Doing all this would require "a completely different way of organizing production—one based not on making a profit but meeting human need."[22]

The institutional framework for this reconfigured society is an open question, and much of the detail would have to be worked out in practice, as experience would show what works and what does not. For example, rather than describe in detail the organizational arrangements of an ecosocialist society, Foster, echoing many others, states only that it would replace the anarchy of the market with comprehensive planning.[23] But some features of the new system are clear. Heavily armed nation-states poised to attack one another for various reasons would have to disappear, replaced by new forms of community. Anything from world government to radical decentralization could replace the sovereign state but, in all cases, the new governing entities would be guided by values of peace and solidarity. The capitalist class would be dispossessed, and management of firms, enterprises, and the economy as a whole would be the job of the "associated producers," workers, and citizens acting democratically and collectively for the common good. In an economy no longer structured by commodity exchange, planning would guide production, including the amount of goods and services produced. Ecosocialism "entails abolishing the capitalist mode of production in favor of a social formation where social (re)production is planned and steered consciously by empowered workers and citizens themselves."[24]

Moreover, revolutionizing the relations of production would enable complete transformation of the technological and organizational apparatus, which "carries in its structure the imprint of its development at the service of capital accumulation and the unlimited expansion of the market."[25] Thus, ecosocialism would not amount to the workers simply taking over the means of production and operating them as before. It would mean comprehensive changes in how goods and services are produced. Above all, this would include how energy is produced and employed in the production process. More broadly, the class relationships embodied in the tools and machinery of production, exemplified by the assembly line and the sweatshop, would be reconfigured to reflect egalitarian social relations in which the health and well-being of the workers and others are paramount.

What would ecosocialism mean for adaptation to climate change? Ecosocialists would claim first that, like the steady state and degrowth, its approach to climate change would result in lower GHG emissions and thus less damage requiring adaptation. Ecosocialism could presumably enable mitigation and adaptation to work in tandem to avert climate catastrophe. However, it is highly unlikely that ecosocialism will be adopted soon, which means capitalism will determine the concentration of GHG in the atmosphere. As we have seen, that means global temperature will rise about 3.0°C above the preindustrial average no later than 2100. Accordingly, if ecosocialism does prevail over the coming century, it will have to find ways to cope with dangerous climate change.

The great advantage of ecosocialism in adaptation would be its capacity for democratic planning. In addition to planning and implementing appropriate infrastructure projects, behavioral changes, and the like to enhance resilience, it would encompass redirecting the entire economy, setting priorities, deciding what and how to produce, and deciding which activities should be curtailed and which expanded.

Reliance on commodity markets to determine what and how much is produced entails a short-term orientation toward immediate gains and lack of coordination between different economic sectors. The remedy for capitalism's "anarchy of the market" is socialist planning that takes full account of the ecological implications of human activity. Presumably, the comprehensive rational plan, worked out in bottom-up consultations, would include provision for effective, equitable adaptation to the unavoidable impacts of climate change. Socialist planning would ensure that adaptation measures would not result in maladaptation, and that the costs of responding to climate change do not fall disproportionately on the poor and developing countries. (Indeed, in a socialist world, there should not be any poor, and perhaps not any *developing countries* as we understand the word today.)

Regarding social organization, one of the more developed socialist proposals is economic democracy. Economic democracy would represent transformational change in the principles of society and in its organization. A global order built on economic democracy would be empowered to undertake appropriate adaptation to the higher temperatures and severe impacts on the ecology and society to come. In addition, the proposal for economic democracy provides answers to two fundamental questions facing the ecosocialist call for transforming global order: where are we going, and how do we get there?

Economic democracy can take a variety of forms, but those focusing on the workplace likely have the most potential for enabling effective adaptation. The growing climate crisis demands immediate action, and emphasizing democratization of the workplace offers a possible, if not probable, path. Rather than seizing the state and directing a transformation of society from the commanding heights, economic democracy calls for change at the point of production and would embody the appropriate form and level of transformational change needed for equitable, effective adaptation. Examining specific proposals indicates why this might be so.

Richard Wolff has made the case for workers self-directed enterprises (WSDEs) to replace the current corporate structure of ownership and management. Wolff disdains efforts to reform capitalism amounting to weak regulations that are discarded when convenient, as well as dictatorial "state capitalism" as represented by Stalinism and the like. A more attractive alternative must be offered, or else calls for transformational change "will not inspire or mobilize a social movement" that workers and citizens see as anything other than another dead end. To inspire change will require "a genuinely different organization of productive enterprises: workers' self-directed enterprises."[26] By contrast to the corporation, the directors of a WSDE would be those working at the enterprise. Outside shareholders would have no control over the enterprise's operations, and the directors would be chosen by the workers, who could also replace them. Consequently, self-directed means workers deciding, in concert with surrounding communities, what, how much, and by what means to produce goods to meet human needs.[27] In a WSDE, those who produce and those who reap the returns are identical, so that no other party exists to manage the enterprise to serve its own interests or take part of the returns.

Wolff analyzes the practical operations and needs of WSDEs operating within a society based on such entities. Periodically, the workers will assemble to consider how to allocate the enterprise's resources, deciding among wages, reinvestment, and services to support the WSDE's operations. The latter includes taxes, clerks, security guards, lawyers, and the like. Thus, the WSDE is not an isolated, self-sufficient economic unit. It is embedded in a social system that includes a government responsible for the provision of public goods. Indeed, Wolff suggests that one of the WSDE's expenses might be lobbying state officials, and the members might turn to public education for their children. However, the needs of the WSDE and the aim of expanding the scope of WSDE enterprises will guide their involvement in public policy and education. To prevent a slide back to capitalism, WSDEs will be concerned to prevent the regular crises besetting a capitalist economy, resulting from overproduction, underconsumption, technological change, and, presumably, approaching ecological limits. Cooperation among WSDEs would be required to manage these problems and ensure equitable treatment of all, as well as encouraging technological progress. Transforming the workplace, Wolff maintains, would affect all other aspects of society, so the WSDE would not simply be an addition to the entities operating within current institutions. If WSDEs did rise to a prominent position, they would need "to change many social conditions to be able to survive, expand, and secure mass support."[28]

Wolff discusses how WSDEs would handle environmental issues,[29] noting that capitalist and state capitalist enterprises have shown little regard for the effects of their choices on the health of workers or the impact of their operations on surrounding communities. The managers of the typical capitalist enterprise live far away from the point of production, and shareholder ownership means the legal owners have almost no awareness of, much less control over, daily operations or investment choices. Thus, "environmental concern is typically a luxury that private and state capitalists believe they cannot afford."[30] By contrast, workers and their families mostly live near the workplace, and they experience directly the effects of the enterprise's operations.

Moreover, WSDEs can incorporate a range of concerns in their policies and decisions, including the environment, which can take priority over the bottom line. An entity with wider concerns for the social, cultural, political, and environmental factors that bear on the well-being of its members would also be likely to insist that government policy give priority to the environment. Adaptation, in particular, would be a high priority for some WSDEs, as climate change might pose direct, severe threats to the enterprise and to its members. A society predominantly made up of WSDEs would engage democratic, cooperative discussions on how to address the threats posed by climate change, doing so at all levels from the local to the global. While there is no guarantee that WSDE deliberations would lead to a given outcome, "we can know that environmental concerns are far more likely to be raised and counted in reaching decisions in a WSDE."[31] As the risks and damages resulting from higher temperatures manifest, WSDEs should raise the priority of adaptation commensurately. Of course, this has already been true in international negotiations and government policies, but the character of adaptation responses should be significantly different in a world of WSDEs. The repeated calls by the Intergovernmental Panel on Climate Change (IPCC) for greater equity, equality, and inclusive decisionmaking could be realized in practice, not facing obstacles in the dominance of profit-seeking corporations controlled by a relative handful of immensely rich people.

David Schweikert offers another take on workplace democracy with implications similar to Wolff's WSDEs.[32] His proposal is to retain markets in goods and services, but to do away with labor markets and financial markets. Labor markets would be replaced by democratically controlled enterprises, and financial markets would give way to public investment.

Schweikert argues that worker-owned enterprises would not face the same incentives as shareholder-owned capitalist firms to compete in a dog-eat-dog market, nor to expand to increase profits. Expansion is essential to the capitalist firm to increase returns to shareholders and to avoid being taken over by competing firms. In a worker-owned enterprise, expansion would only mean more people sharing the product, income, and profits with no great change in the return to each individual worker, so the owners and managers would have little incentive to grow. And there would be nobody to sell out the firm to outside bidders, so the pressures of corporate predation would not exist. Enhanced productivity from adopting new technology could be translated into reduced work hours rather than increased output. The option of nonmaterial incentives, such as reduced work hours, would be essential to avoiding the environmental harms of unbridled growth.[33]

Schweikert also suggests significant change in finance and investment. In a capitalist system, investment is guided by investors' perception of profit opportunities. By contrast, "a sustainable democratic alternative to capitalism must also democratize financial markets."[34] Public investment financed by a flat tax on all enterprises would replace private financial markets, with investment funds allocated according to a formula, with due exceptions for systemwide needs and special circumstances. Schweikert does not mention it, but presumably exceptions would be made for communities and regions facing severe climate impacts and needing significant adaptation funding.

In sum, ecosocialism takes a step beyond the steady state and degrowth in suggesting alternatives to capitalism. The steady state is, by some accounts, compatible with capitalism, and degrowth advocates vary from those who believe degrowth can work in a reformed capitalism to those who believe it requires socialism. Ecosocialists are firm in declaring that capitalism and ecological health are incompatible and a new social, economic, and political order must replace it to avoid transgressing planetary boundaries. In light of the failure of state-led socialism in the Soviet Union and other countries, contemporary ecosocialists are averse to recommending such a seizure of the means of production for management by an authoritarian state. Rather, they believe democracy is the key to transforming society toward sustainability, and the main practical concern is to undertake and implement comprehensive planning rather than leaving the future to the uncertainties of the market. While most ecosocialists choose not to go much beyond those propositions, some forward more specific plans such as economic democracy in worker-owned enterprises. The spread of such entities is expected to influence the rest of society toward values and practices supporting workplace democracy.

If the process of change tends toward ecosocialist values, the broad outcome might be ecological civilization (EC). This would mean that global movements to defend the environment, workers, human rights, and communities will have succeeded in saving civilization.

Ecological Civilization

The proposal to create an ecological civilization has been discussed at least since the 1980s. Then, Soviet scientists advanced the idea, and the Chinese scholar Ye Qianji adopted it and helped raise its profile in Chinese political discourse. By 2007, EC had become part of the Chinese government's political guidelines, promoted as the successor to industrial civilization.[35] The main task of EC is to raise awareness that humans and civilization are embedded in a global ecosystem with its own dynamics, and this awareness would allow the world to avoid the pitfalls of barbarity and decadence. The main barrier is self-regarding individualism, which subverts the culture and education that would provide the foundation for EC. Transitioning to EC will demand broadly accepted worldviews with "a clear vision to aspire beyond shopping."[36]

Fred Magdoff addresses the need for an ecological worldview by showing how principles of ecology can inform the construction of an EC.[37] Magdoff claims the science of ecology indicates that an ecological civilization should display self-regulation through democratic decisionmaking and worker control over enterprises; diversity of opinions, talents, and opportunities; ecological efficiency so that waste and exploitation of resources is minimized; self-sufficiency in essentials such as food; and resiliency through self-renewal. The last point is especially pertinent to adaptation, for Magdoff asserts that self-regulation, self-sufficiency, diversity, and ecological efficiency contribute to creating a resilient society. A resilient society is a society that adapts well to disturbances, including the impacts of climate change. Consequently, Magdoff's recommendation of a transition to ecological society is a call to create a society with ample capacity to adapt to the risks and harms of climate change. Undoubtedly, an EC would also manage energy production and consumption differently to achieve significant mitigation. But some impacts are already unavoidable, requiring fundamental transformation of society in the direction of EC to meet successfully. The principles, organization, and decisionmaking processes of EC would presumably enable equitable, effective adaptation.

According to Foster, an ecological civilization would be an aspect of the revolutionary transformation from capitalism to socialism.[38] Foster and others call for the transcendence of class-based civilization in which the domination of nature and the domination of human beings work in tandem to perpetuate a system of class rule and privilege. The worsening ecological crisis points to the objective historical necessity for an ecological revolution and an ecological civilization, leading to global ecological socialism. Thus, EC would entail transformations of the technology and social relations of production, governance, and culture, yielding "a social formation aimed at substantive equality and ecological sustainability, emphasizing human development."[39]

Ecological civilization would be a civilization that stays within the safe operating limits of the planet, and it would do so without a sudden population crash, abandonment of major cities, reverting to peasant agriculture, forgetting about and dismantling modern technology and knowledge, or, broadly, giving up on modern civilization altogether. Rather, Magdoff avers, civilization would grow only to the point of meeting human needs, enhance cooperation and sharing, and "operate with respect for, and care of, the environment—locally, regionally, and globally."[40] Democracy and substantive equality would be essential to achieving and sustaining an ecological civilization.

Importantly, the idea of EC is not to recreate precapitalist civilizations, say by reverting to class societies in which a privileged elite enjoys opulence by appropriating the agricultural surplus. Nor is it to move toward some form of precivilization society of small-scale agriculture or hunter-gatherer subsistence. "I may be forgiven," writes Joel Kovel, "for insisting that New York, Paris, London, and Tokyo not be taken down in an ecological society."[41] To the contrary, he claims, an ecological society would maintain such large-scale activities as railroad systems, communications systems, and power grids. Rather, an ecological civilization would save technologically advanced civilization, but with a strong element of sustainability and harmony with nature.

What would ecological civilization mean for adaptation to climate change? Similarly to ecosocialism, EC would represent transformational change, which is one aspect of the IPCC's understanding of adaptation. Transformational change provides the institutional framework within which equitable, effective, efficient, and just adaptation policies and projects can be implemented. The fundamental principles of EC, say its advocates, do not present barriers to taking needed steps, in contrast to the principles of global capitalism. This would be true regardless of the level at which global temperature stabilizes. If the UN Framework Convention on Climate Change (UNFCCC) negotiations and national policies show some progress, then perhaps the equilibrium temperature will be below 3°C, as analysts now expect, and an EC would take the steps needed to protect communities across the planet from the risks and impacts both anticipated and experienced at that level. Whereas global capitalism would respond to the challenge by prioritizing capital accumulation, EC would prioritize human well-being for all. But could EC come into being in time?[42]

The question of how much time remains raises the problem of transition from current arrangements to an ecological civilization.[43] The same question presents itself regarding the steady state, degrowth, and ecosocialism, albeit with varying degrees of urgency. Ecological civilization requires the broadest and deepest transformation of global society, for which at present the political will does not exist, and the inertia and power of global capitalism appears too great to transcend before catastrophic temperature increases occur. If the transition to EC occurs over decades or centuries after global capitalism has already locked in 3°C temperature rise or more in the next century, the magnitude of the crisis might be too great for civilization of any kind to address.

Addressing the temporal question, Chris Williams claims that the transition to an ecological civilization will be quicker and less violent than the centuries-long transition to capitalism.[44] Rather than the long process of gathering together scattered production in decentralized farms and villages, EC will be erected on capitalist property, which is already largely socialized. Capitalists, relatively few in number, have used every instrument at their command to transform the possessions of the great mass of the people into capitalist private property. By contrast, the transition to ecosocialism or ecological civilization would mean that the great mass of the people would convert capitalist private property into the common possession of those who are engaged directly and indirectly in production.

But even if Williams is correct, this provides little guidance on how the transformation is to occur in the short time remaining to prepare for a much hotter world, much less one in which tipping points have been breached and uncontrollable climate change, lasting for centuries, prevails. Some analysts find hope in a global movement to enact ecological values rooted in Indigenous peoples, women, the poor and marginalized, and a possible environmental proletariat.[45] This broad global social movement for ecology has been termed *Blockadia* in light of its potential capacity to prevent environmentally harmful development.[46] These tendencies exist now, so advocates of a transition to EC can point to capacities in the world today that could push toward such transformational change.

Against this, skeptics would cast doubt on the ability of an environmental social movement to affect policies and practices at the core of capitalist development. Mounting a citizen campaign to block a pipeline or a coal mine is a far cry from transforming the social order.

And experience shows that an immense amount of effort goes into organizing people for ecological ends only to see investors and corporations prevail with end runs around citizen activism or evasion of successful movement outcomes. For example, years of heretofore successful efforts to block a pipeline running from West Virginia to Virginia have been subverted by a deal between the Joe Biden administration and Senator Joe Manchin (D-WV). Cases like this can be multiplied endlessly. Worse, as movements gain power, the deployment of power in support of capitalism will become more forceful. The Virginia pipeline case can be fought out in courts and legislatures, but a real threat to the rule of capital would likely result in the mobilization of police and military power to crush the movement. As Arran Gare observes, "The dominant powers do not tolerate any significant delinking from this system that would threaten their power."[47] This is not to say organizing and action to bring about change is hopeless or futile. It is to recognize that instituting global EC is no easy matter, and the movement to do so must think in terms of decades of struggle across many fronts. The world might not have that much time.

Ecological civilization, ecosocialism, degrowth, and, to a lesser extent, the steady state all raise similar questions of feasibility. They share a critique of growth-oriented global capitalism, differing somewhat on whether growth is inherent and essential to capitalism. But in practice, if not in theory, growth and capitalism have gone hand in hand. Consequently, all these perspectives suggest a new world order that does not require growth and that orients economic activity toward meeting basic human needs. This forces consideration of two vital concerns: where we are going, and how to get there. This book has suggested that economic democracy provides a viable answer to those questions. Others see some hints for where and how to achieve ecological civilization in the Chinese experience. China has made EC part of its plan for the development of the country, even including the goal in its constitution.

As a country built consciously on socialist principles and aspirations, embracing a Marxist analysis of capitalism and society more broadly, perhaps China shows the promise of achieving EC. In China, asserts Foster, "revolutionary-scale ecological reforms are being attempted even in a context of rapid economic growth aimed at bringing China up to a level with the West."[48] Foster claims China is making a good faith effort to establish EC on a rapid timetable, as shown by its willingness to give up some economic growth to pursue ecological goals. This effort, Foster states, enjoys broad popular support as well as the endorsement of top government officials; grassroots environmental action is a powerful force that points the way toward a novel form of ecological communism. Certainly, by comparison to other countries, Foster writes, China is advanced in enacting ecological values: "This is largely due to its role as a post-revolutionary, socialist-oriented social formation that retains a large element of economic planning capability, state direction, and collective values, invigorated by continual popular mobilization in both rural and urban areas."[49]

Gare adds to this that leaders, scholars, and citizens in the West should support EC, which would enhance China's global leadership. China's success in becoming a leading world power, Gare claims, would help to move the entire world toward a new kind of civilization built on diversity and common values.[50] This would enable replacement of today's commitment to self-regarding, endless expansion of the production of commodities and view of life as a struggle for the survival of the fittest with an ecological standpoint, encompassing such concepts as James Lovelock's Gaia hypothesis.[51] Gare and others see China's turn toward EC as, in part, a recovery of ancient Chinese values emphasizing harmony with nature,[52] while Foster asserts China's EC represents a step in Marxist analysis and practice not requiring reference to ancient traditions. Either way, China's global leadership toward EC is commendable and worthy of support.

Not everyone agrees with this assessment. The Western voices supporting China's EC take note of continuing environmental problems in China, but they consider these challenges as temporary and understandable in light of China's starting point as a fast-developing country in a world of established capitalist states. Yet Paul G. Harris (who has lived in Hong Kong for decades and published numerous books and articles on environmental issues) asserts that, contrary to the view that China is on its way to a benign future, its embrace of EC is simply a ploy to ensure the continuation of authoritarian rule.[53] According to Harris, China's incorporation of EC into its constitution, and the leadership's frequent invocation of its commitment to a future EC, is "greenwashing" on an epic scale. Harris points out that China's interpretation of EC is that it is a balancing of environmental and commercial interests, not that commercial interests would take second place or be supplanted altogether—an interpretation clearly at odds with that of Foster and others on the Western left. Thus, Chinese EC is about making economic growth less harmful, far from the completely transformed priorities and institutions that advocates of EC expect.

Moreover, at present, commercial concerns dominate, so most of China's environmental indicators are moving in the wrong direction, including that the country remains the largest source of GHG emissions. Rather than creating a sustainable society, China's policies are a form of coercive environmentalism that strengthens the state party complex, which leaders assert is essential to bringing about EC. Consequently, the official line is that to oppose the authorities is also to oppose a healthy environment. Succinctly, Harris comments that China is the world's greatest environmental threat, and current trends show nothing to alleviate that problem. China's embrace of EC, he asserts, is to justify more authoritarian rule, so that "Instead of being a solution to China's burden on the environment, ecological civilisation reinforces party-state dominance and growing material consumption and pollution."[54]

While declaring itself the champion of ecological civilization, China remains one of the most environmentally harmful societies on the planet and, Harris concludes, *ecological* is not a term aptly applied to Chinese civilization. To be sure, the country's leadership has forwarded EC as a long-term aspiration, not a description of its current situation, so perhaps it should be judged according to its intentions rather than current actions. Still, the differences of perspective regarding Chinese EC suggest that China is not yet a model for the future to be emulated without serious critique and modification.

The advantage of tying the project of EC to China is that a powerful country can provide the vehicle for social change as nothing else can. For all the hopes invested in the environmental proletariat and Blockadia, it is doubtful that such movements can overcome the entrenched power of capitalism to bring about fundamental social change. However, if joined by the power of China, social movements in the rest of the world could make headway. To take an analogous situation, while the USSR and other socialist bloc countries existed, hopes for a possible transition to socialism and an end to the rule of capital remained viable, even if on life support. The demise of the Soviet experiment led nearly everyone to agree with Margaret Thatcher that "there is no alternative" to capitalism. With the fall of the Soviet bloc, there was no national champion for workers movements. Likewise, without a powerful state to support the cause, prospects for the ecology movement dim considerably. However, were China to take seriously its role as world leader on the path toward EC, with dozens of developing countries and perhaps some of the richer and developed countries joining in, the world might see a movement toward a society in which the values, worldviews, and practices compatible with true sustainability could be realized. Thus, a lot is at stake regarding the sincerity of China's commitment to EC.

It is useful to note in this connection that words sometimes do matter, and cynical ploys to hold on to power can get away from their perpetrators. When President Ronald Reagan called for democracy in Latin America, it was widely seen as a bad-faith gesture intended to rob opposition movements of popular support while boosting the legitimacy of pro-US dictatorships. Countries such as El Salvador were to stage "demonstration elections" that would amount to public relations for brutal oligarchies so US interests could be defended behind a facade of democracy. But within two decades, Latin America did indeed hold meaningful elections across the region, in countries that had never seen them before, and to restore democratic governments where they had been lost due to US involvement such as in Chile. Given the opportunity to cast ballots, the people of Latin America have elected progressive and even socialist governments, and Latin American democracy, it turns out, is not a cover for sustained US imperialism. Similarly, cynical Chinese party functionaries seeking to perpetuate authoritarian rule under the guise of environmentalism might find that the project escapes their control. If so, then EC in China could become the genuine model for social change that its supporters desire. What begins as a cynical ploy can become a world-changing process. However, some might consider this a thin reed on which to place the planet's future.

Moving from today's world order dominated by neoliberal capitalism to some combination of the steady state, degrowth, ecosocialism, and EC would constitute transformational change. Society would be built on a new set of principles, institutions, and priorities. Although the exact outlines of the new society are unclear, perhaps the "abstract utopian" vision that ecosocialists and others advocate will inspire global movements to get there. If so, then this transformed ecological civilization will be able to allocate resources and promulgate policies that can provide equitable, effective adaptation measures. However, the challenges of feasibility and time stand in the way of such a world-changing transformation. Some observers doubt that it can or will happen, which means the world will remain on the path to ecological catastrophe. According to this view, the best option is to prepare to rebuild some form of civilized society after the collapse of today's global civilization.

Notes

1. Damian F. White, Alan P. Rudy, and Brian J. Gareau, *Environments, Natures and Social Theory: Towards a Critical Hybridity* (London: Palgrave Macmillan, 2016), p. 52, emphasis in original.

2. Herman Daly, *Steady-State Economics: Second Edition with New Essays* (Washington, DC: Island Press, 1991).
3. Herman Daly, "Toward Some Operational Principles of Sustainable Development," *Ecological Economics* 2, no. 1 (1990): 1–6.
4. Robert Costanza, Gar Alperovitz, Herman Daly, Joshua Farley, Carol Franco, Tim Jackson, Idą Kubiszewski, Juliet Schor, and Peter Victor, *Building a Sustainable and Desirable Economy-in-Society-in-Nature* (Canberra: Australian National University Press, 2013), p. 17.
5. Robert Costanza, Gar Alperovitz, Herman Daly, Joshua Farley, Carol Franco, Tim Jackson, Idą Kubiszewski, Juliet Schor, and Peter Victor, *Building a Sustainable and Desirable Economy-in-Society-in-Nature* (Canberra: Australian National University, 2013), p. 73.
6. Mark Lynas, *Our Final Warning: Six Degrees of Climate Emergency* (London: 4th Estate, 2020), pp. 275–278.
7. Robert Costanza, Gar Alperovitz, Herman Daly, Joshua Farley, Carol Franco, Tim Jackson, Idą Kubiszewski, Juliet Schor, and Peter Victor, *Building a Sustainable and Desirable Economy-in-Society-in-Nature* (Canberra: Australian National University Press, 2013), p. 53.
8. Robert Costanza, Joshua Farley, and Ida Kubiszewski, "Adapting Institutions for Life in a Full World," in *State of the World 2010*, edited by Erik Assadourian (Washington, DC: Worldwatch, 2010), pp. 85–90.
9. Francois Schneider, Joan Martinez-Alier, and Giorgio Kallis, "Sustainable Degrowth," *Journal of Industrial Ecology* 15, no. 5 (2011): 654–656.
10. Serge Latouche, "Degrowth Economics," *Le Monde Diplomatique,* November 2004.
11. Jason Hickel, "What Does Degrowth Mean? A Few Points of Clarification," *Globalizations* 18, no. 7 (2020): 1105–1111.
12. Jason Hickel, "The Anti-Colonial Politics of Degrowth," *Political Geography* 88 (2021), p. 102404.
13. Jason Hickel, "What Does Degrowth Mean? A Few Points of Clarification," Globalizations 18, no. 7 (2021), pp. 1108–1110.
14. Jason W. Moore, *Capitalism in the Web of Life: Ecology and the Accumulation of Capital* (London: Verso, 2015).
15. John Bellamy Foster, "Capitalism and Degrowth: An Impossibility," *Monthly Review* 62, no. 8 (2011): 26–33.
16. Jason Hickel, "What Does Degrowth Mean? A Few Points of Clarification," *Globalizations* 18, no. 7 (2020), p. 1106.
17. Ying Chen, "Degrowth—What's in a Name? Assessing Degrowth's Political Implications," *Monthly Review* 75, no. 3 (2023): 160–170.
18. Serge LaTouche, *Essays on Frugal Abundance* (Westerville, Ohio: Simplicity Institute, 2014), https://simplicityinstitute.org, p. 21.
19. Michael Lowy, "Nine Theses on Ecosocialist Degrowth," *Monthly Review* 75, no. 3 (2023), p. 156.
20. Michael Lowy, "Nine Theses on Ecosocialist Degrowth," *Monthly Review* 75, no. 3 (2023).
21. Chris Gilbert, "'Where Danger Lies . . .' The Communal Alternative in Venezuela," *Monthly Review* 75, no. 3 (2023): 72–84.
22. Chris Williams, *Ecology and Socialism* (Chicago: Haymarket Books, 2010), pp. 168–169.
23. John Bellamy Foster, "Planned Degrowth: Ecosocialism and Sustainable Human Development," *Monthly Review* 75, no. 3 (2023): 1–29.

24. Guney Isikara and Ozgur Narin, "Degrowth and Socialism: Notes on Some Critical Junctures," *Monthly Review* 75 no. 3 (2023), p. 34.
25. Michael Lowy, *Ecosocialism: A Radical Alternative to Capitalist Catastrophe* (Chicago: Haymarket Books, 2005), p. 22.
26. Richard Wolff, *Democracy at Work: A Cure for Capitalism* (Chicago: Haymarket Books, 2012), p. 102.
27. Richard Wolff, *Democracy at Work: A Cure for Capitalism* (Chicago: Haymarket Books, 2012), p. 103.
28. Richard Wolff, *Democracy at Work: A Cure for Capitalism* (Chicago: Haymarket Books, 2012), p. 112.
29. Richard Wolff, *Democracy at Work: A Cure for Capitalism* (Chicago: Haymarket Books, 2012), p. 117.
30. Richard Wolff, *Democracy at Work: A Cure for Capitalism* (Chicago: Haymarket Books, 2012), p. 135.
31. Richard Wolff, *Democracy at Work: A Cure for Capitalism* (Chicago: Haymarket Books, 2012), p. 135.
32. David Schweickart, *After Capitalism* (Lanham, MD: Rowman and Littlefield, 2002).
33. David Schweickart, *After Capitalism* (Lanham, MD: Rowman and Littlefield, 2002), pp. 113–123.
34. David Schweickart, "Is Sustainable Capitalism Possible?" *Procedia Social and Behavioral Sciences* 2, no. 5 (2010), p. 6745.
35. Aran Gare, "Barbarity, Civilization and Decadence: Meeting the Challenge of Creating an Ecological Civilization," in *Yearbook of Philosophy in Process,* edited by Michel Weber and Ronny Desmet (Louvain-la-Neuve, Belgium: Presses Universitaires de Louvain, 2009), pp. 167–189.
36. Aran Gare, "Barbarity, Civilization and Decadence: Meeting the Challenge of Creating an Ecological Civilization," in *Yearbook of Philosophy in Process,* edited by Michel Weber and Ronny Desmet (Louvain-la-Neuve, Belgium: Presses Universitaires de Louvain, 2009), p. 181.
37. Fred Magdoff, "Ecological Civilization," *Monthly Review* 62 no. 8 (2011): 1–25.
38. John Bellamy Foster, "The Earth-System Crisis and Ecological Civilization: A Marxian View," *International Critical Thought* 7, no. 4 (2017): 439–458.
39. John Bellamy Foster, "The Earth-System Crisis and Ecological Civilization: A Marxian View," *International Critical Thought* 7, no. 4 (2017), p. 450.
40. Fred Magdoff, "Harmony and Ecological Civilization: Beyond the Capitalist Alienation of Nature," *Monthly Review* 64, no. 2 (2012), p. 6.
41. Joel Kovel, *The Enemy of Nature: The End of Capitalism or the End of the World?* (London: Zed Books, 2007), p. 185.
42. For an incisive analysis of the difficulties facing the transition to ecosocialism, see Michael J. Albert, "Ecosocialism for Realists: Transitions, Trade-Offs, and Authoritarian Dangers," *Capitalism Nature Socialism* 34, no. 1 (2023): 11–30. On the need for a rapid transition to meet the Paris goals, see Kevin Anderson, John F. Broderick, and Isak Stoddard, "A Factor of Two: How the Mitigation Plans of 'Climate Progressive' Nations Fall Far Short of Paris-Compliant Pathways," *Climate Policy* 20, no. 10 (2020): 1290–1304.
43. Gerard Ahearn, "Towards an Ecological Civilization: A Gramscian Strategy for a New Political Subject," *Cosmos and History* 9, no. 1 (2013): 317–326.
44. Chris Williams, "How Will We Get to an Ecological Civilization?" *Climate and Capitalism,* November 2, 2015.

45. John Bellamy Foster, "The Earth-System Crisis and Ecological Civilization: A Marxian View," *International Critical Thought* 7, no. 4 (2017): 439–458.

46. Naomi Klein, *This Changes Everything: Capitalism vs the Climate* (New York: Simon and Schuster, 2014), pp. 293–336.

47. Arran Gare, "After Neoliberalism: From Eco-Marxism to Ecological Civilization, Part 2," *Capitalism Nature Socialism* 32, no. 3 (2021), p. 53.

48. John Bellamy Foster, "Ecological Civilization, Ecological Revolution," *Monthly Review* 74, no. 5 (2022), p. 5.

49. John Bellamy Foster, "Ecological Civilization, Ecological Revolution," *Monthly Review* 74, no. 5 (2022), p. 7.

50. Arran Gare, "From 'Sustainable Development' to 'Ecological Civilization': Winning the War for Survival," *Cosmos and History* 13, no. 3 (2017): 130–153.

51. James Lovelock, *The Vanishing Face of Gaia* (New York: Basic Books, 2009).

52. Mary Evelyn Tucker, "The Challenge of Creating Ecological Civilization in China," Yale School of the Environment, 2017, https://environment.yale.edu.

53. Paul G. Harris, "China's 'Ecological Civilisation'—A Large-Scale Form of Greenwashing?" *Hong Kong Free Press,* July 23, 2022.

54. Paul G. Harris, "China's 'Ecological Civilisation'—A Large-Scale Form of Greenwashing?" *Hong Kong Free Press,* July 23, 2022.

7

Life on Hothouse Earth

MANY PROMINENT VOICES IN THE CLIMATE CHANGE DIScourse have pointed to the possibility of a difficult or even uninhabitable Earth if humanity stays on its current path.[1] Bill McKibben writes of a "tough new planet" in which increased engagement in climate politics will require "building the kind of communities and economies that can withstand what's coming."[2] While McKibben insists the struggle for mitigation must continue and become stronger, he also indicates that the probable inadequacies of mitigation will require transformed communities and societies adapted to a hot world. As the crisis worsens, the possibility of successful adaptation recedes and the degree of transformation needed increases, not a hopeful circumstance. Due to climate change, he writes, "The habitable planet has literally begun to shrink, a novel development that will be the great story of our century."[3] John Michael Greer identifies several sources of civilizational decline, including climate change and other ecological limits. He writes, "Civilizations normally leave a damaged environment behind them when they fall, and ours shows every sign of following that wearily familiar pattern."[4] James Lovelock claims, "We are in a fool's climate . . . and before this century is over, billions of us will die and the few breeding pairs of people that survive will be in the arctic region where the climate remains tolerable."[5] In the same vein, climate scientists have investigated the planet's habitability at plausible temperature increases occurring over several centuries, identifying limits to human survival.[6] More extreme than these, James Hansen has suggested the possibility, albeit remote, of a planet as hot as Venus unless large reductions in greenhouse gases (GHG) emissions occur.[7] Obviously, such a world would be beyond any amount of adaptation. Long before that day would come, however, humans would have to adapt in many ways to extend the life of the species and society; it is unlikely civilization would survive until the final days.

The upshot of such speculation, backed by dire warnings from some climate scientists,[8] is that civilization might be incompatible with ecological health in general and adaptation to climate change in particular. Consequently, the various ways by which civilization might be salvaged, from ecological modernization to ecological civilization, are probably futile. The die is cast, and human civilization has already lost the game. In sum, Rupert Read asserts, "This civilization is finished."[9] What, then, is to be done when profound transformation of human society will be imposed as the ecological crisis, led by global warming, worsens?

Deep Adaptation

Recently, the prospect of the collapse of civilization has led scholars and activists to the idea of Deep Adaptation, a way of coping as the institutions of modern civilization crumble, and rebuilding after the collapse. Deep Adaptation is about human survival when irreversible failures of the various systems sustaining modern society have ended civilization. Read comments that it is time to consider what will follow civilization, and to begin engaging in Deep Adaptation, attempting to slow the current destructive process, which means that "dramatically courageous things are now necessary."[10]

Read identifies three possible futures. First, the world could make a smooth transition to some form of ecological civilization. Second, we could plan for and implement the transition to a postcollapse civilization. Third, we could do nothing, but failure to act would mean total ecological and civilizational collapse, leading to possible human extinction. Although the first might be the most desirable, the time is too short and the obstacles too large for it to be a realistic option. The third option would be to surrender and to underestimate human capacity for response to even the worst eventualities. Thus, the second is our "best hope." Consequently, "we need to think about what comes after the likely collapse of this civilization and plan accordingly." Read suggests that planning for a postcollapse successor civilization requires awareness of the magnitude of the crisis; collective deliberation to turn despair into action; serious inquiry into the nature of a new civilization; and construction of strong communities able to carry as many people as possible through an extreme crisis to an era of constructing a new civilization. Most pertinent to this book, Read suggests, "We need to take adaptation preparation seriously, as well as deepen and transform our concept of it."[11]

Advocates of Deep Adaptation and similar orientations to climate change paint a bleak picture of the future.[12] Current policy, they say, would not prevent a cataclysm even if the commitments made in international negotiations were honored. Unceasing economic growth, climate tipping points, and misguided hopes for geoengineering ensure that the future will be hotter than the temperature thresholds adopted at the Paris conference and in the Conference of the Parties (COP) process since. Climate targets will not be achieved: "This means," asserts Read, "that unprecedentedly dangerous climate change is coming and it is going to get a lot worse for a long time to come, accelerating broader ecological degradation."[13] Barring rapid, sweeping changes in the global order, life will become precarious. While dire warnings might once have seemed abstract and distant, Jem Bendell avers that "starvation, destruction, migration, disease and war" will occur in the current generation's lifetime, and "you will fear being violently killed before starving to death."[14] Thus, Deep Adaptation is remarkable for its insistence that civilization might collapse within decades; the signs of collapse are already visible.[15] Whereas numerous observers assert the world has only a short time to avert 2°C or more global average temperature increase, Deep Adaptation claims the same urgency to prepare for life after civilization.

Worse, change is coming faster than climate science had expected. Today's news is reporting that the possible collapse of the Atlantic Meridional Overturning Circulation could occur in this century, disrupting climate patterns worldwide.[16] As noted, 2023 was the hottest year on record; climate experts say such extremes of rising temperatures will be the new normal, not a mere fluctuation of weather. In the summer of 2023, Phoenix, Arizona, suffered a full month of high temperatures over 110°F, as heat waves hit the US Southwest and South, as well as other countries around the world. Iran, Iraq, and South Asia sweltered in unprecedented heat waves. South America experienced 100°F and higher temperatures during the Southern Hemisphere's winter. Meanwhile, the oceans surrounding Florida had become so warm that biologists resorted to removing coral from the sea and putting it in onshore tanks to save it.[17] Meanwhile, despite the increasing use of renewables, the amount of oil consumed in the post-Covid-19 years has reached record highs,[18] and the world oil market is expected to continue to grow.[19] Such evidence of rapid change in the climate and lack of effective action motivates the Deep Adaptation call for planning now in preparation for a possible collapse in this century.

Yet as this book has indicated, much of the research on climate change broadly, and adaptation in particular, is oriented toward management of problems occurring at local and regional scales. Most of it assumes the likely success of mitigation to hold temperatures low enough for the impacts to be manageable, while a considerable share simply ignores the dire consequences of weak mitigation; namely, a world in which we cannot adapt within current institutional frameworks. Accordingly, Bendell observes that "the field of climate adaptation is oriented around ways to maintain our current societies as they face manageable [climatic] perturbations."[20] Thus, existing scholarly literature gives little attention to the prospect of near-term societal collapse. Deep Adaptation, like other perspectives reviewed in this book, challenges these managerial approaches to climate change and calls on motivated citizens to adapt by preparing for the imminent collapse of major systems supporting civilization.

Deep Adaptation asserts that little can be done at this stage to avert catastrophic impacts of climate change on the ecology and human society. This is not because of a lack of policy and technological alternatives, but primarily because it is highly unlikely that current institutions will adopt the measures needed to avert catastrophe or cope with the drastically changed climate that business as usual will have created.[21] Committed to sustaining economic growth, today's global governance is not suited to embracing the radical, transformative measures that would be required to perpetuate civilization.[22] Instead, "the deep adaptation perspective sees the pace and scale of dangerous levels of climate change and ecological destruction to be so fast that neither a reform of capitalism nor of modern society is realistic."[23] Indeed, modern civilization is caught in a dilemma: continuing on as before portends ecological collapse, but preventing ecological collapse might require inducing a social and economic collapse. Consequently, "it has become difficult to imagine a future for our civilization."[24]

Creating a new world order that would preserve civilization and avert the coming ecological catastrophe would require improbable revolutionary change, leading Read to note, "For it to have any chance of success, it would require the speedy overcoming of virtually all the vast vested interests as well as of ignorance, apathy and lethargy, amongst the other forces that stand in the way."[25] This suggests that ecosocialists and those hoping for a new ecological civilization are delusional. If the climate crisis is severe enough to force such a transition, it is also too profound and rapid to expect ecological civilization to arrive before the collapse. Thus, with civilizational failure all but certain, as many people as possible must overcome their psychological resistance to contemplating the end of civilization, meaning the end of their accustomed ways of life.

Contrary to the fear that painting such a bleak picture will lead to apathy and denial,[26] advocates claim research supports the Deep Adaptation assertion that facing the truth is empowering.[27] Only confronting the reality of the coming collapse will equip people with the attitudes and orientations required to prepare for a postcollapse reconstruction.

In this respect, the notion of the "system of interest" is significantly different from that posited in Intergovernmental Panel on Climate Change (IPCC) reports and most scholarship on climate change. For the mainstream, the system of interest would be existing institutional and social arrangements and practices, and adaptation is about what can sustain some version of civilization. By contrast, for Deep Adaptation, the system of interest does not yet exist, and its attributes are unknown. It is whatever social system emerges after current arrangements have been swept away; Deep Adaptation is not in any way about saving the current global order, which is assumed to already be doomed. Without Deep Adaptation's eye to the future, there might be no social systems at all, hence no system of interest to adapt. But it remains that, for Deep Adaptation, the system of interest is an imagined world of uncertain dimensions. Adaptation, then, is about preserving and creating the conditions that would enable the construction of a new civilization. Clearly, this is far removed from the understanding of adaptation one finds in official reports or in most scholarship, in which the language of coming catastrophe is often a preface to tepid recommendations for reformist policy responses.[28]

With the world heading toward near certain catastrophe, exactly what is to be done now to ensure the possibility of a new civilization? Deep Adaptation recommends a redefined notion of resilience, along with relinquishment, restoration, and reconciliation. Resilience for Deep Adaptation means to survive with certain worthwhile norms and behaviors. Thus, rather than referring to a system's capacity to return to its previous state after a shock, Deep Adaptation defines *resilience* as surviving the shock with the capacity to create a new society better suited to the radically changed conditions of a much hotter and ecologically stressed world. *Relinquishment* entails letting go of those assets, behaviors, and beliefs that make matters worse such as vulnerable urban areas and communities, harmful industries, and consumerism. *Restoration* "involves people and communities rediscovering attitudes and approaches to life and organisation that our hydrocarbon-fuelled civilisation eroded."[29] *Reconciliation* calls on people to recognize their common fate and to avoid causing more damage by acting out of fear of one another.

In addition, Deep Adaptation calls for *relocalization* in anticipation of the collapse of global economic arrangements.[30] Communities cannot rely on imports of food, for instance, when complex communications, transportation, and financial systems have broken down. Governments (whatever form they take) might have to exert a strong hand in allocating food, say by imposing drastic reductions in meat consumption. Energy production (renewable, of course) will likely be decentralized, and regulations will cease favoring large corporations operating vast energy grids. In addition, governments and communities might have to set up alternative payment mechanisms, withdrawing from private financial markets. Organizations such as the World Bank and the International Monetary Fund will serve no purpose and will disappear. Importantly, communities will need to offer mental health counseling and mutual support to help people cope with the sense of loss, fear, and grief likely to attend civilizational collapse. Efforts to create alternative, intentional communities that are self-sufficient and more likely to survive the demise of centralized economic and political institutions have been under way for many years, and Deep Adaptation advocates cite them as examples of what can be done.

However, the weight of existing national and global institutions makes for a difficult context for such initiatives. "A political power centre in its dying throes might 'redistribute' any and all wealth towards itself,"[31] write Matthew Slater and Skeena Rathor. One might add that the elites in business and government will use any means necessary, not excluding world war, to ensure upward redistribution continues. Consequently, if Deep Adaptation is about facing the truth, that truth must not stop at acknowledging the ecological catastrophe to come. It must also face the truth about how dying empires behave. The story is not hopeful for alternatives based on localized organic farming and the like, facing wealthy, heavily armed parties seeking to maintain their privileges and power. Nor is it clear that local governance would be superior to broader national and regional institutions. In practice, local governments can be manipulated by self-interested parties,[32] as well as lacking in the resources needed to carry out essential adaptation programs.

This is not to say that Deep Adaptation has entirely missed the mark. No doubt, the system of states, multinational corporations, and global financial institutions, along with their police forces and militaries, present formidable barriers to fundamental change. But the story that Deep Adaptation tells would have these massive structures falling of their own weight in an ecological crisis of their own making. Post-civilization communities only have to keep their collective head down so as to survive an epic crisis and begin rebuilding after collapse. They would not have to fight against social, economic, and political arrangements that will have died anyway. In addition, the local communities of the postcivilization world would not be akin to municipal governments in the United States. They would reflect entirely different principles and occupy an entirely different natural and social environment. Such pitfalls as the capture of the city council by real estate interests would be irrelevant. Nevertheless, simply surviving civilizational collapse as coherent communities will be challenging. Rather than peaceful, intentional communities striving to build ecological civilization on the ruins of today's collapsed civilization, the world might see roving bands of plunderers taking advantage of the relative weakness of the carriers of Deep Adaptation.[33] That might not spell human extinction, but it would mean that planning for a postcollapse rebuilding project could be futile.

Hospice Earth?

Whether the ecological crisis will be severe enough to threaten the survival of civilization is unknowable today. If it is, then the outcome will have been ensured long before the extent of the damage will be visible, so there will be no time to remedy the problem. Accordingly, the abiding matter of the relationship between mitigation and adaptation shapes answers to the pressing questions on whether this civilization is finished. If international negotiations result in effective mitigation, holding global average temperature rise low enough to avoid dangerous climate change, then perhaps civilization can survive and even thrive. If not, then the future looks considerably more doubtful. Katharine Hayhoe succinctly poses the alternatives: "It turns out that for our modern world, the difference between a higher versus a lower emissions future is nothing less than the survival of our civilization. In the lower emissions scenario, our agriculture, our water, and our economic systems can continue, albeit with significant and often costly adaptations. The higher emissions scenario predicts the end of many of these systems as we know them."[34]

Accordingly, difficulty arises in placing hopes in an international system and governments that have heretofore shown limited capacity to meet the challenge. Certainly, preventing every tenth of a degree of warming matters, so whatever can be achieved in UN Framework Convention on Climate Change (UNFCCC) negotiations, national policies, and private sector action will help. Nevertheless, the path is toward higher temperatures, much higher than the Paris goal of 1.5°C or less.[35] Consequently, nothing less than the survival of civilization as we know it is at stake.[36]

Climate scientists mostly share the view that a world of 4°C higher global average temperature is incompatible with the civilization that emerged and developed during the Holocene, and it remains possible that the world will see such high temperatures. After all, the 2°C threshold for policy was accepted because the science suggested that, beyond that level, dangerous tipping points would have been reached, meaning humans could not do much to halt further warming, so higher temperatures would be seen eventually. Under current policy and practice, the likely temperature increase in 2100 will approach 3°C, at which point the threat of surpassing tipping points increases dramatically. While 3°C is considerably below previous expectations of 4°C by 2100, it is still well into dangerous territory, and warming would continue beyond that policy horizon. This suggests that, assuming current commitments are met, the temperature rise to 4°C and higher will have been delayed, not prevented. If mitigation commitments are breached and cast aside in a security or economic crisis, little will have resulted from decades of climate negotiations, not even a delay in the hour of reckoning. Moreover, today's experience with climate change impacts indicates that severe harms occur at lower temperatures than previously thought, so it may be that delaying the temperature rise does not mean delaying the social crisis. Thus, within a few decades, the world will know whether civilization has a chance. Beyond that, it will know whether it is possible to adapt at all, deeply or not.

Hayhoe's observation points to a broader concern about research and policy on adaptation. That is that the developed world, including the most populated of the fifty US states, is vulnerable and possibly lacking the resilience to adapt even minimally to climate change, never mind achieving climate-resilient development (CRD). Indeed, the more complex and dependent on integrated infrastructure, commerce, and regulated markets a society is, the more vulnerable it would likely be to high temperatures. Those systems sustaining such a society would break down. Societies with less reliance on complex energy, transportation, and industrial systems, perhaps more oriented toward most people living on small farms and selling goods in local markets, might be in a better position to adapt to dangerous climate change.

This upends the more or less implicit framing of adaptation discourse, with its legacy of colonial attitudes and assumption of the benefits of modernization. Sybille Bauriedl and Detlef Muller-Mahn cogently observe that discourse on climate change evokes vulnerability striking hardest in tropical areas deeply affected by European colonialism, such that predominant views of the vulnerable bear the marks of that experience. In official and scholarly documents, the victims of climate change are poor, they live near the equator in underdeveloped countries, and they are mostly women. "Adaptation policy," Bauriedl and Muller-Mahn remark, "is absolutely focused on rural and indigenous communities living in this region. And these regionalised adaptation discourses are dominantly formed by scientific knowledge produced by scholars of the Global North."[37]

This mainstream view of adaptation collapses when the problem comes home to developed societies, people other than poor women living in the tropics. And the collapse of the developmentalist, modernization-oriented adaptation discourse signals the bankruptcy of the wider view of strength, resilience, and capacity in a North that must come to the rescue of a weak, vulnerable, and incapable South. Indeed, more than a few observers believe that it will one day be necessary for the women, Indigenous peoples, and peasants of the Global South to rescue the people suffering the failures of industrial, modernized civilization. Thus, thinking through adaptation to climate change can overturn the implicit assumptions, prejudices, and moral judgments of the current global order. This includes the developmentalist orientations of policymakers in the Global South as much as those of the presumed dominant elites of the Global North.

The question that could confront people alive today (not only future generations) is whether the *planet as we know it* is at risk, which is what the concept of planetary boundaries addresses. Does the advent of the Anthropocene mean the safe operating environment for humanity will soon be lost? Is it already too late to avert that outcome? If not, how much time remains? If it is too late, what is to be done? Such questions loom over all approaches to adaptation to climate change, even those that refuse to acknowledge pessimistic answers.

Deep Adaptation goes far in providing negative answers to such questions. The conditions congenial to human civilization, North and South, are at risk and likely to be lost soon. Not enough time remains to transform global social, economic, cultural, and political practices toward genuine sustainability. Most official policy amounts to little more than debating the dinner menu on *Titanic* while reinforcing existing hierarchies regarding who is allowed at the table. Thus, civilization is finished, and the responsible course of action is to prepare to build a new one. Deep Adaptation's answer to lifeboat ethics is to ensure there are plenty of lifeboats and their crews are working together to reach a safe harbor rather than raiding each other for scarce supplies.

However, not even Deep Adaptation takes the analysis far enough into the political realm to provide a realistic assessment of the future under such conditions. This civilization might be finished but it is not dead, it refuses to accept its coming demise, and it will fight with every weapon at its disposal until the final day comes. Thus, rather than midwife a new civilization, the survivors might witness a long process of civilizational death, one that sweeps away societies large and small no matter their aspirations to rebuild. The apt metaphor for that agonizing process is ameliorative care for the terminally ill. In short, for those lucky enough to survive somewhere removed from the main scene of conflict, their best option might be to prepare to survive on Hospice Earth, where little more can be done than to ameliorate the pain of death and ease the transition to their collective demise.[38]

Assessing Adaptation

We have seen that adaptation is increasingly recognized as necessary; the meanings of adaptation and transformational change are contested; and, contrary to the view that adaptation offers an easy way out of ecological crisis, adaptation implies instead that global institutions might require radical transformation. However, the prospects for radical transformation are dim.

It is likely that the time for implementing mitigation sufficient to prevent dangerous temperature rise has already passed. If so, the choice is not between transformed institutions or making do with current institutions. The choice is between having adaptation imposed under dire circumstances or making proper provision for the future. The former will likely entail bitter conflict as societies collapse in the face of increasingly difficult climatic conditions. The latter offers some hope that a transition can be made to equitable, peaceful social arrangements adjusted to Hothouse Earth. Failing that will leave us in Hospice Earth.

"For ecological collapse threatens the survival, if not of our species, then at least of the current social order," asserts Sven Lütticken.[39]

This examination of adaptation to climate change shows the necessity of engaging profound questions of human social organization to cope with a warming world. Adaptation is no technical, managerial matter. Without committing itself to a well-defined vision of the future, the IPCC has insisted that meeting the challenge could require far-reaching social and economic change based on principles of inclusion, equity, and justice, principles that are in short supply in today's world. While the IPCC has been criticized for going beyond its mandate to assess policy options and undertake policy advocacy,[40] even the most managerial elements of adaptation policy engage deeper questions of social justice. It is unavoidable to speak about adaptation while ignoring the institutional and normative issues that adaptation raises. Moreover, attempting to confine the discussion to supposedly technical issues is itself to take a normative stand on the appropriate principles and institutional arrangements for adaptation policy; namely, to endorse the status quo, including to assume, perhaps mistakenly, its viability. For its part, the IPCC remains well within the mainstream of policy discourse, suggesting climate-resilient development as the appropriate response; CRD seeks to combine mitigation with sustainable development, implying some form of green growth. Clearly, the range of options extends far beyond that, and which path the world takes will result from political struggle. Political disputes will never yield to technocratic remedies, as scientists and policymakers quickly learned when they developed the thinking on adaptation.

Adaptation was discussed in the earliest prominent scientific conferences on climate change. Indeed, in the 1970s, scientists tended to assume that observed trends in energy use would continue uninterrupted and adaptation to changing climate conditions would be the main response. But examining such topics as the adaptation of agriculture, urban areas, and coastal regions forced consideration of the political and normative dimensions of adaptation. It was increasingly evident that adaptation was far more a political and social matter than one of natural science or technical management. Soon enough, this realization manifested in the debates over climate policy in UN processes, particularly in the UNFCCC conferences. While some special interests advanced the notion that adaptation would allow unrestricted consumption of fossil fuels to continue, this cynical version of adaptation did not gain much traction. Instead, by far the predominant concern was that developing countries would be facing the greatest impacts with the most meager resources. Thus, developing countries pushed adaptation onto the global agenda and kept it there for decades, until at last the Paris conference in 2015 placed adaptation firmly on the same level as mitigation. However, setting an international agenda is not the same as action on the ground, and adaptation remains sporadic and underfunded around the world.

In its Fifth Assessment Report (AR5), the IPCC brought the concept of transformational change to the center of the climate policy dialogue, then elaborated it further as CRD in its Sixth Assessment Report (AR6). Unfortunately, one person's transformation is another person's incremental change. Leaders, policy analysts, activists, and scholars have taken up the banner of transformation without any consensus on what the word means. Even for the radical voices, specifying institutional arrangements has been avoided in favor of abstractions and general principles to guide a new society, and for good reason: the more detailed the prescription for the future, the more obvious the barriers to transition become. Taking such barriers seriously can lead to despair, or it can lead to hopes for a new society emerging from the ashes of the old. It does not suggest a path for salvaging current global institutions. Many observers agree, "We will have to create something different and better or risk collapse into something far worse."[41] But it is difficult to say what would be better or how to get there.

Nevertheless, millions of people take to the streets, attend meetings, sign petitions, vote, and express opinions urging action on climate change. Every COP draws tens of thousands of climate activists, demanding that this time—this time—the various officials in attendance take climate change seriously. The officials often respond with rhetoric that would sound familiar at an Extinction Rebellion (XR) rally, wholeheartedly agreeing that time is short, today's opportunity cannot be squandered, the world must act now. But the official product invariably falls well short of solving the problem. Instead, governments implement policies directly contrary to their urgent pledges at the COPs, such as providing huge subsidies to the fossil fuel industry.[42] The futility of pressuring global neoliberal capitalism and a heavily armed system of sovereign states to modify their priorities and practices enough to mitigate or adapt to climate change has led to such proposals as Deep Adaptation.

However, the global movement to bring about an effective response to the climate crisis has had no more than incremental effects, in part due to the inertia of dominant institutions, and in part because agreeing on the destination has proven difficult. Deep Adaptation does not engage the people who take to the streets, being primarily an intellectual current rather than a mass movement. Extinction Rebellion and related movements do mobilize large numbers, but XR does so while avoiding political programs so as to unite the broadest possible coalition. Indeed, such unity as can be achieved to demand action on climate change exists because such important terms as *transformation* are open to a wide range of interpretations. Moreover, empirical support for the effects of environmental movements on government policy is mixed, leaving aside bringing about sweeping social transformation. A recent study notes that "the evidence remains unclear as to whether democratic incrementalism embodied in existing practices will be sufficient to generate the sustainability transformation at a scale needed to avoid irreparable disruptions in socio-ecological relationships and systems."[43]

This book has documented that adaptation was at first viewed as a technical problem, one to be addressed within existing institutions. The long journey to today's understanding of adaptation has brought us to notions of adaptation as a response to the collapse of existing institutions. Within that spectrum, one finds calls for institutional reform and radical change. For its part, the IPCC has recently moved toward CRD to reconcile the conflicting and contradictory tendencies attending adaptation. What is to be made of these many voices and perspectives?

In sum, whether it is the mainstream's muddling through, ecosocialism's planned transition to a world of social justice, Deep Adaptation's postcollapse reconstruction, or a new Dark Age, adaptation will occur in some form. Which form it will take beyond current incremental adaptation is unknowable because it depends on how much mitigation occurs, how severe the impacts are on human society, how much political will is mustered to implement effective adaptation, and how big the level of support is for bringing about social transformation in the midst of a potentially worsening climate crisis. The probable path is toward global temperature averaging well above the consensus on what constitutes dangerous climate change, with little sign of adequate preparation for that tough new planet. This leaves the world in an uncomfortable position: the necessary is not feasible, but the feasible is not sufficient.

Theory, Policy, and Future Research

Chapter 1 reviewed various theoretical approaches to international relations and environmental politics that could illuminate the politics of adaptation to climate change. This book has described an outcome in need of explanation: a general record of weak responses, missed opportunities, and responsibilities shunned. In part, the inadequacy of mitigation accounts for the rise of adaptation on the global agenda, but the response to adaptation needs has been no better. Although some theories would draw attention to moments of progress, the general story is one of failure. Taking this depiction of the outcome as roughly accurate, how does the record of adaptation politics and policy bear on theoretical debates? What does it suggest for future research directions?

Theory

Recall that realism would expect that the struggle of nation-states for survival and power would overwhelm environmental concerns, and adaptation would be a minor issue on the global agenda. Liberal institutionalists posit a more varied global agenda, a diverse set of actors, and more opportunities for cooperation on common problems, putting considerable emphasis on the origins, changes, and effect of international regimes and global governance. Constructivism challenges the materialism of realism and liberalism, asserting that the essential question is how actors create their social worlds in language and interaction.

Green theory ranges across a wide spectrum. On one end, green theory encompasses game theory analysis suggesting that the tragedy of the commons will lead to a "lifeboat ethic" in which the more powerful must throw the less powerful off the boat to save themselves. Other rational choice analyses are more optimistic, asserting the tragedy can be averted through cooperation that yields the best outcome for the world, overcoming short-sighted individual incentives that would lead to collective failures.[44] Green theory also emphasizes the influence of nonstate actors on environmental politics and policy, moving attention away from the state-centric focus of mainstream international relations theory. Green theory in general insists that ecology is more than another issue area alongside security and political economy, to be understood in familiar frameworks developed well before the environment rose on the global agenda. The advent of the Anthropocene has introduced novel aspects of social life requiring new tools of analysis. Ecosocialists assert that the world faces an epochal crisis of capitalism requiring comprehensive social transformation.

This book's study of adaptation to climate change offers some purchase on these abiding theoretical debates. A realist would not be surprised by the anemic response to mitigation and adaptation responses to climate change that has been observed and would see little cause for optimism about the future. Security concerns drive states, primarily fear of vulnerability to military attacks by other states. To ensure their own survival and to obtain some degree of leverage over events, states elevate enhanced national power to the top priority. In today's world, economic strength is an element of national power. States will not want to lose relative standing as economic powers because to do so would weaken some of their sources of influence, and it would lessen their capacity to build and maintain their military.

This has several significant environmental consequences. Military spending absorbs a large share of global output that could be directed toward such purposes as adapting to climate change. Strong economic growth will be embraced as a way to enhance national power, adding to the imperatives for growth outlined earlier. Militaries wreak considerable environmental damage in everyday operations as well as in armed conflict. The drive for security pushes other priorities down the list and fosters short-term thinking. Great-power war would sweep away all environmental agreements and delay efforts to mitigate and adapt to a hotter world. For realists, all this is to be expected in an anarchic world dominated by heavily armed, power-seeking states. Thus, it is no surprise that climate change policy has resulted in plenty of symbolic displays of concern with little substance resulting. Nor would a realist expect things to get any better. Even as the ecological crisis worsens, states will be seeking to maintain their relative status, regardless of the collective result.

Consequently, realism can claim validation from the record of adaptation politics and policy. The attention to adaptation has risen, and the rhetorical orientation to adaptation has changed significantly toward the need for transformation to achieve adequate *adaptation* (variously defined). But actual policy and financial commitments are paltry, with little sign this situation will change. National governments might well increase efforts within their own domains on experiencing negative impacts, but the inadequacies of current and past international adaptation efforts are highly likely to persist; rich countries' assistance to poor countries will never be adequate to fund adaptation, not to speak of compensation for unavoidable damages.

However, observations that line up with a given theory are not, in themselves, evidence that the theory is correct. Otherwise, any noncooperative outcome in international affairs would validate realism. Making the case requires providing evidence that the patterns and dynamics posited by realism were, in fact, responsible for the observed outcomes. In this regard, realists could cite the unhelpful influence of several great powers, including the United States and Russia. The United States was a leader in the early days of climate negotiations, but it shifted to become a laggard within the decade.[45] Depending on which party controls the White House, the United States has been either a drag on climate progress or an outright saboteur. The failure of cooperation when one or more major powers stand in the way is a straightforward expectation of realist theory. Presumably, the United States and others see effective, equitable adaptation policy as inimical to their national interests in maintaining relative power, sustaining the economic growth that is such a large part of the power struggle, and wielding a capable military. To be sure, the US military itself considers climate change to be a serious national security threat and it has programs to adapt to climate change.[46] But that is about ensuring that climate change does not harm national power and security, and it involves the self-regarding policies of the state, with international cooperation incidental to the main purpose.

Further, realists must also cope with other aspects of adaptation politics and policy outlined in this book. For one, international negotiations on adaptation have occurred, showing some, albeit inadequate, progress. For the most part, adaptation will occur locally, and it will be funded out of local and national resources. So why should it be on the international agenda at all? Why does it play such a significant part in international climate negotiations? While power, security, and the national interests are undoubtedly part of the story of adaptation, perhaps other factors are in play that explain what might be seen as anomalous in realist theory.

Liberal institutionalism would begin to respond by observing that progress has occurred on climate change, even though the sovereign state system presents numerous obstacles. Liberal institutionalism expresses optimism about the capabilities and prospects for current global arrangements, with due expectation that they will change in the right way. If adaptation to climate change becomes a top priority on the international agenda, states and other actors will create new regimes and modify existing ones to produce appropriate norms and rules to guide behavior. Despite the obstacles, states understand that the national interests include concerns besides military security, threats to national well-being can arise from sources other than states, and rational apprehension of the long-term common interest in addressing the climate challenge enables cooperation on mitigation and adaptation. Working through international organizations such as the UNFCCC, states have cooperated to change the trajectory of GHG emissions. No doubt, much remains to be done, but without cooperative action, the path of business as usual would be toward the high end of temperature outcomes, whereas the current path is well below that. In short, the situation is not good, but it could be a lot worse.

Similarly, the international community has gone from paying little attention to climate adaptation to establishing a Global Goal on Adaptation. Outcomes in this area are due to a variety of factors. The agenda is diverse, some opportunities for international cooperation have been seized, and progress on adaptation is visible. Where policy falls short, the best explanation remains liberal institutionalism: failure can result from lack of sufficient understanding of the problem, lack of salient solutions, and the unhelpful influence of nonstate actors. That is, success and failure hinge on much more than the distribution of power across sovereign states.

One question for liberal institutionalism would be whether an international regime for adaptation to climate change exists. If so, it would provide important evidence for liberal internationalist theory. The term *international regime* is defined as the norms, rules, principles, and procedures around which actor expectations converge in a given issue area. A strong regime affects behavior, leading actors away from suboptimal outcomes resulting from self-interested action, toward cooperative behaviors resulting in collectively optimal outcomes. In such areas as trade, this logic has been well developed: norms of liberal trade policy reject tariff barriers and conform to principles of trade theory such as comparative advantage. Thus, states enact rules in the World Trade Organization and other institutions to regulate open trade, and these rules are enforced by penalties against violators of trade agreements. These add up to a significant global rejection of trade protectionism and embrace of free-trade policy; that is, an international trade regime. Does a similar institutional arrangement exist in regard to adaptation?

The general norms of international environmental politics embodied in international law and agreements bear on the adaptation issue area, including obligations to future generations, common but differentiated responsibilities (CBDR), polluter pays, the precautionary principle, and loss and damage (L&D), all guided by the objective of climate-resilient development.[47] The review of adaptation negotiations indicates that these norms are only partially incorporated in commitments and behavior. Adaptation efforts are often reactive rather than anticipatory, leaving the interests of future generations only sporadically represented, despite much rhetoric at COPs and elsewhere about the mandate to do so. Common but Differentiated Responsibilities regarding adaptation has been largely discarded as developed countries have rejected the assertion that they should bear the burden of adaptation in other countries. By the same token, polluter pays and L&D have been repudiated. In both cases, to adhere to them comprehensively and thoroughly would mean past polluters accept full financial responsibility for the harms done by recent and future climate change. Instead, responsibility for L&D has been recast from liability to another form of adaptation funding, and, more generally, funding for adaptation is an addition to traditional development programs. Thus, in the normative terms declared in international environmental agreements, a regime for adaptation would be weak, at best, and possibly nonexistent. The operative norms governing adaptation are that most costs will be paid locally with help from national governments, and any contribution from the wealthy countries to the poor countries is voluntary assistance, not required but offered from charitable sentiments. However, this is as much as to say that independent national decisions guide adaptation behavior, which is not the makings of an international regime. Lacking the norms of an international regime, the rest of the elements are irrelevant. Consequently, in the issue area of adaptation, liberal institutionalism would have to explain failure, not the successful construction of international institutions. While liberal institutional theory does offer explanations for failure, the absence of regimes is hardly compelling evidence for the theory writ large.

From a constructivist perspective, the central question is how states define themselves and others, and how this results from a process of social interaction. Having analyzed that question, the balance-of-power and game theory models that are the stuff of realism and liberalism appear as secondary questions, useful only after the important puzzles have already been solved. Presumably, actors (including states) could redefine their identities, social worlds, and definitions of others in ecologically sustainable ways. If such redefinitions take hold across the planet, behaviors will follow, including those that support ecological sustainability.

By this account, one might note that states did not understand their role to include being stewards of the global environment until well after World War II. The advent of the environment as a national and international concern in the 1960s led to new ideas about state interests and related ethical obligations. Recognition of the varying sources and effects of environmental problems gave rise to such principles as CBDR, polluter pays, and obligations to future generations. With the passage of time, environmental responsibility has become institutionalized as part of national and international practices, no longer questioned as such but taken for granted as what states do and who they are. Thus, nearly every government has some kind of environmental ministry or department, if only because having one is part of being a modern state.[48]

Of course, much room remains for debating the terms of the state's role as environmental steward, and the mechanisms by which that role will be enacted. Understandings of obligations, rights, interests, and identities have changed and will continue to change. One could argue that the social construction of adaptation discourse has lately encompassed the notion of transformational change, destabilizing prevailing understandings of actor identity and appropriate forms of interaction. No doubt, facing an existential threat from unchecked global warming, states seek to preserve their identities as the main actors in world affairs by negotiating policies to address the crisis. But others, observing that even the most well-intended products of international negotiations fall well short of the urgency of the situation, advocate different understandings of the actors and appropriate behavior in a radically changed world. Instead of being a state responsibility for promulgating, funding, and implementing adaptation policies, projects, and programs, for such perspectives as Deep Adaptation, adaptation is seen as completely transforming the institutional landscape, perhaps replacing states with bioregional communities better suited to coping with a dangerous legacy of global warming.

But just as liberal institutionalism must confront the weakness of the international norms relevant to adaptation, so constructivism must recognize how deeply embedded the practices of sovereignty and the national interest are. Constructivism insists that cognitive understandings of identity, appropriateness, and the conditions for action become so taken for granted that action occurs without reflection on how this particular set of actors, with these particular identities, guided by a given set of ideas about appropriate behavior came into being. In fact, the settled ideas and attitudes shaping contemporary climate adaptation policy are not transformational. To the contrary, they are highly resistant to transformation, and often enough are reactionary instead. Consequently, even weak commitments to deal with the climate challenge hang by the thread of partisan politics in the United States, Great Britain, Australia, and the rest.

In sum, major schools of thought in international relations theory vary regarding expectations, with realism anticipating policy failure while liberal institutionalism and constructivism hold out some hope for progress. But they would all have to account for a disappointing record of sporadic responses, inadequate funding, refusal to accept responsibility, and a large dose of self-regarding behavior.

Ecosocialism and related orientations would also survey a landscape of failed policies and weak responses, but they would explain that sorry record in terms of the economic forces driving state policy, corporate behavior, and class relations, as well as how they relate to gender, race, and ethnicity. Capitalism's growth imperative and its quantification and commodification of everything it touches bar the path to equitable sustainability. As a system of class domination, the kind of inclusive, just adaptation that even the IPCC declares is required for a successful response to the climate challenge is impossible. And the conflictual nature of the global capitalist system threatens to sweep away any positive movement. Istvan Meszaros summarizes, "The violent assertion of the destructive imperatives of global hegemonic imperialism, through the formerly unimaginable destructive might of the United States as the global hegemon, cannot bring *global solutions* to our aggravating problems but only *global disaster*."[49] Far from an efficient economic system, capitalism is destructive and wasteful, counting cluster bombs and Barbie dolls as commensurate additions to gross national product (GDP). Such a system, ecosocialism suggests, is highly unlikely to adopt optimal adaptation measures.

However, as noted, adaptation per se has not been much discussed in ecosocialist writing. Assessments of National Adaptation Plans, local adaptation strategies, the responses of small farmers, and similar topics well-represented in adaptation scholarship are not at all discussed by ecosocialist writers. It is not only that ecosocialists tend to speak in broad terms; to the contrary, socialists and ecosocialists are not at all averse to discussing particular topics in current affairs. Commentary on policies for economic management, labor relations, human rights, women's rights, and indeed on GHG mitigation is ample, whereas adaptation to climate change is barely mentioned. Perhaps with the UNFCCC adoption of a Global Goal on Adaptation, ecosocialists will pay more attention to the topic. But adaptation has not so far counted among the current issue areas drawing attention from ecosocialists.

At an abstract level, the relevance of ecosocialist theory is on adaptation as transformational change. The record of missed opportunities, deflections, policy failures, and denial of responsibility outlined in this book would be no surprise to ecosocialist theory. The predominant influence of market values and the commodification of all aspects of social life would render adaptation a low priority, readily swept aside when profit opportunities arise or when broad economic management requires greater concern for economic growth and investment. Thus, the remedy is to transform global neoliberalism into some form of ecological civilization. "In order to create an economically viable, and also on a long-term basis historically sustainable, social reproductive order it is necessary to radically alter the self-contradictory inner determinations of the established one, which impose the ruthless submission of human need and use to the alienating necessity of capital expansion," writes Mezsaros.[50] The new social form would presumably be much better positioned to make the right choices about how to defend coastal cities, cope with floods and droughts, protect people from extreme heat, ensure agricultural productivity, and all the other specific measures that will be required to implement incremental and transformational adaptations to climate change.

Ecosocialism as theory would expect tepid adaptation policy, just as the world has seen. It would locate the causes for that weak response in the economic imperatives driving markets and state behavior. Unlike realism's self-regarding power seekers, states in ecosocialist theory are manifestations of the rule of capital. This is not to say states are necessarily tools of, say, fossil fuel corporations. Rather, the state tends to the broad, common concerns of all firms, looking to shore up markets and class relations so as to perpetuate the system's accumulation of capital; any given company might thrive or go bankrupt, as markets decide. Nevertheless, states will be disinclined to adopt policies or institutional changes that hinder growth and accumulation, meaning the general inadequacy of adaptation.

So, what actually happened here? Did self-regarding states block effective climate action to enhance their relative power? Did the institutions of climate policy lack the necessary teeth, a situation that could be remedied? Did the identities and institutionalized practices of state sovereignty allow some degree of redefinition of identity and purpose while also preventing creative alternatives that would better serve human well-being? Did economic interests and class relationships compel policymakers at all levels to conform climate adaptation policy to the requisites of capital accumulation?

The easy answer is all of the above. States, institutions, ideas, and classes all matter for adaptation politics and policy. So does the growing perception that a serious social crisis could ensue from rising global temperatures, one requiring incremental adaptations, transformational adaptations, and perhaps transformational change. This book began with the hunch that adaptation policy, even as it rose on the global agenda, had become tamed; rather than implying profound changes in the global order at all levels, it instead would have been folded into the traditional development discourse so that existing hierarchies, roles, and privileges were not threatened. For the most part, that hunch has proven accurate. But gaining more from the investigation requires explaining which theoretical perspective provides the most leverage on this important issue.

Class relationships shape all the various responses to climate change, from incremental and transformational adaptation to transformational change and Deep Adaptation. In some cases, the class basis for society is acknowledged and incorporated into the analysis, while in others it is the implicit and often unquestioned background condition for deeming a proposal "realistic" and feasible.

A good place to begin an analysis from a class perspective is to recall comments regarding Jason Moore's analysis of capitalism in the web of life. Moore, correctly, observes that the same quest for cheap inputs into the production process has driven capitalism for centuries. Moore identifies four cheaps (labor, natural resources, food, and energy) necessary to maintain profitability, or, in Marxist terms, the extraction of surplus value. This imperative to hold input costs down stands behind such phenomena as dispossession of peasants, colonialism, child labor, and the mobilization of women. But Moore asserts, capitalism has reached its limit, in that the capacity of the ecosystem broadly understood provides few opportunities to reduce costs. Climate change is a significant barrier to obtaining cheap energy from fossil fuels, it will raise the costs of production of food, and natural resources required for such things as electric vehicles will become more expensive. Consequently, capitalists will see few opportunities for accumulation by appropriation, which are required for capitalist profitability.[51]

A few comments are in order on this account of capitalist development. To begin, to suggest that nonhuman entities—resources, food, energy, and the like—are sources of value runs against the key insight of a Marxist analysis of capitalism. Karl Marx posited a market system in which all commodities trade at their value, including raw materials and labor. Regarding objects, such as loads of coal or bales of cotton, the purchaser must pay full value, which means there is no possibility of earning a profit (extracting surplus value) from those inputs. Only labor, said Marx, has the unique attribute of producing more value than it embodies itself. Consequently, when Moore suggests that natural resources and the like are sources of surplus value (profit), he is implying that commodities other than labor sell below full value. Recognizing this departure, Moore turns to the history of capitalism rather than the abstract analysis of the somewhat idealized economy that Marx analyzed.[52] Marx explicitly simplified the model so as to reveal the inner workings of capital. Stepping into the historical realm means that the complications of real life continue to operate, which implies that capitalists might be extracting profits from nonlabor inputs. In actual practice, profit is only partly gained at the point of production in the form of surplus value extracted from labor; it is also gained from acquiring nonhuman inputs below their value and reaping the gains by selling finished products as commodities at full value. That is, value results from exploitation of labor and appropriation (taking it by one means or another).

Is this a sensible move? It may well be historically accurate to portray the accumulation of capital as part appropriation, part exploitation. However, the important distinction between a barrel of oil and a human being remains. The barrel of oil, whether cheap or expensive, does not care how it is acquired, exchanged, or used. It has no interest in the distribution of the returns from its use. Its well-being does not depend on receiving fair value for its contribution to the production process. The same is true of all the other nonhuman commodities and resources entering into the production process. Labor is the only commodity that produces more value than it embodies but, more importantly, it is the only participant in the production process that cares about the question. Workers, their families, and their communities are the direct opposite of the barrel of oil on all counts. To argue that nature, in some way, contributes value in the same way labor does can muddy the picture and obscure the profoundly political character of capitalist production.

Further, one might ask why do these particular cheaps make the list and not others? Why do food and energy, but not housing? One might say that these items are closer to nature than others, but is a thatched hut further from nature than a tanker truck full of gasoline, or a field of genetically modified organism (GMO) wheat? It would seem not. It would make more sense to account for all the inputs into the production process—the oil, the bales of cotton, the raw sugar, the iron, the plywood, and so on—as commodities employed in the production of commodities, as well as commodities for final consumption. This is how a firm accounts for its costs. It appears there is nothing special about three of the cheaps. But it remains that the fourth cheap, labor, is quite special.

The critical moment in this arrives when opportunities for profiting from appropriation are insufficient to avert a direct confrontation between the two great classes. No doubt, in the historical development of capitalism over the course of five or more centuries, the quest for cheap inputs has driven capital all over the globe. In one place, it found cheap labor, in another cheap land, in another cheap agricultural products. However, as noted above, if Moore is correct about this, and if he is also correct that capitalism is reaching its limits, it must be that the fields of opportunity to reap these gains from appropriation are running out. It must be that there is not much cheap energy, food, or natural resources left to be had. In other words, it must be that all those inputs will trade at full value; accumulation by appropriation will be at an end, or at least sufficiently exhausted that it offers no remedy to the problems the system confronts. That, again, implies that only one input to the production process remains that can provide a profit—labor. In sum, capitalism will have become a system based on the exploitation of labor alone, and it cannot soften the consequent direct conflict of interest by reducing the costs of other inputs. And that means capitalism will have reached the end of its capacity to adapt in myriad ways to sustain the process of capital accumulation, all those ways that do not require a direct confrontation with the only other party to the process that cares about the outcome—workers and all who depend on them.

Adaptation to climate change, along with responses to all the other ecological and social crises that global capitalism has created for itself, would be revealed as a matter of political power. Matthew T. Huber, in a study of class and climate, writes that focusing on conflict at the point of production reveals that the source of the problem is "a small minority of owners who control and profit from the production of the energy, food, materials, and infrastructure society needs to function, and the key to change is 'confronting that small minority of owners.'"[53] This is a useful recommendation for analyzing the politics of adaptation to climate change. However, by itself, focusing attention on the relations of production does not explain capitalism's remarkable ability to adapt to changing ecosocial conditions or when that capacity will have reached its end.

Moore's analysis reveals that the hard limit of adaptation arrives when accumulation by appropriation is no longer sufficient to maintain profitability. The direct confrontation that Huber emphasizes of workers versus owners has been evaded for centuries as capital has located, invented, developed, and appropriated cheap sources of energy, food, and other requisites for production and for maintaining the labor force. Workers have had every reason to support these efforts; if the alternative was a massive, inevitably violent struggle for socialism on one hand or descent into barbarism on the other, it made sense for the workers to prefer to avoid the crisis. But workers have cause to confront capital when capital can no longer evade the crisis of its own making. Consequently, it is only when an unavoidable crisis looms that the struggle will be joined in earnest. As temperatures rise, costs will rise as well, blocking capital's usual avenues for acquiring cheap inputs and evading the crisis. In this light, the question of adaptation to climate change is momentous.

The Policy of Hope

Scientists in early discussions could ignore the broader political implications of adaptation because it still appeared to them to be a technical and managerial problem well within the capabilities of experts to address effectively. As the dangers of unchecked global warming became more widely recognized, the political, economic, and social character of adaptation became clear. Much of the debate turned on the financial responsibility of the Global North for adaptation in the Global South. Adaptation politics and policy were primarily framed in terms of nation-states and the goals of the development agenda. Despite the IPCC's emphasis on the possible need for transformational adaptation toward CRD, the IPCC and the scholarship on which its reports rely could not or would not confront how the dynamics of capital accumulation limited options and blocked equitable, effective, inclusive adaptation. For the most part, the mainstream response was to call for restructured markets that would incentivize energy efficiency and reduce externalities, along with adequate public and private investments in adaptation projects such as early warning systems, improved irrigation, and stronger infrastructure. No doubt, the IPCC, drawing on a growing body of scholarship and reflecting the voices of climate activists, has insisted that adaptation could mandate transformation of "current development pathways," but it has avoided specifics about those pathways, rarely mentioning capitalism, much less the class relations that define it. Accordingly, the recommendations following from this influential form of analysis of adaptation are mostly incremental adaptation measures, and, where necessary, more extensive transformational adaptation (moving communities and the like). Transformational change is, for all intents and purposes, off the global agenda.

To be sure, the IPCC might have adopted the only "realistic" approach to the problem. The prospects for transformation to the steady state, degrowth, ecosocialism, or ecological civilization in time to mitigate global warming, or in time to develop the institutional framework for equitable, inclusive, effective incremental and transformational adaptation, are low. The only path forward, then, is to achieve as much mitigation as possible, hope that there are no devastating tipping points, hope that the impacts of climate change are not too dire, and hope that existing institutions will take at least enough action to preserve civilization.[54] If such hopes come true, then today's global and national economic, social, political, and cultural arrangements can continue on their way. Ideally, public funding along with markets and corporate investments, encouraged by appropriate policies and incentives, will lift billions out of poverty while managing transitions in every economic sector—energy above all—and providing support for loss and damage due to climate change as well as anticipatory adaptation to more loss and damage to come.[55]

Yet relying on so much hope to save civilization is a dangerous gamble. Briefly describing the hopeful future shaping UNFCCC negotiations, IPCC reports, national policies, and much of the scholarly and activist literature might suggest that this vision is utopian. Indeed, the hopeful view still acknowledges that global average temperature rise will reach the danger zone by century's end, just not as high as once thought. Hope, it seems, lies in the thought that global average temperature in 2100 will rise to only a potentially catastrophic level rather than a certainly catastrophic level. After 2100, all bets are off.

From an ecosocialist perspective, capital's demands for profit and accumulation would block most of this hopeful agenda anyway, including everything from incremental, reactive adaptation to transformational, anticipatory adaptation. Consequently, the only option is to remove the barrier to rational climate policy—the rule of the capitalist class—and replace it with management by the associated producers; that is, the workers. But if reforming capitalism to meet the challenge is unlikely, overthrowing it before the damage is permanently done is even more remote. Thus, the Deep Adaptation perspective that civilization is finished might be correct—if the impact of climate change is catastrophic and such adaptation as can be achieved is insufficient. Even so, the Deep Adaptation proposal to shelter in place until the worst of the storm passes and build something new afterward faces some problems with reality as well. As noted, the chaotic postcollapse world might be quite hostile to the small, intentional, and green communities that are supposed to carry the world to a successor civilization. Most or all could well be wiped out or subjugated by heavily armed private armies of plunderers. Unfortunately, it may well be that such rootless bands are the best adapted to the postcollapse world that Deep Adaptation anticipates.

Consequently, hope could amount to complacency and denial. This more or less willful blindness, writes Marco D'Eramo, encompasses "the overwhelming majority of the world's political class, irrespective of ideological orientation—who feign concern from their air-conditioned offices and private planes, and then do nothing." Indeed, he continues, they offer misguided half-measures that only promise to perpetuate social practices that created the problem.[56]

Meanwhile, far from feigning concern, the Heritage Foundation has produced a blueprint for the Republican Party to employ in reversing nearly every aspect of US climate policy. The report asserts, "Mischaracterizing the state of our environment generally and the actual harms reasonably attributable to climate change specifically is a favored tool that the Left uses to scare the American public into accepting their ineffective, liberty-crushing regulations, diminished private property rights, and exorbitant costs."[57] Supposedly, the lying left thus stands in the way of sensible production of fossil fuels, along with a variety of other economic pursuits, and a reformed federal bureaucracy would remove these impediments to growth. The general recommendation is that controlling pollution, including GHG emissions, should only occur if it does not cost very much or inconvenience anyone. This blueprint may well guide the next administration if catastrophe compounds catastrophe in climate policy and an administration hostile to climate policy takes office. So much for hope.

Future Research

Scholarly work about adaptation to climate change is growing rapidly.[58] One hesitates to suggest that scholars should pay attention to this or that aspect of the issue when pertinent studies may well be published or under way by the time this book reaches readers. In lieu of a detailed survey of the literature, perhaps a few general remarks will suffice.

To begin, one must frankly recognize that under some conditions, speaking of future research needs can be ludicrous. If the dire predictions of civilizational collapse are realized, it would mean that further scholarly research would barely exist, much less have any use. If complex institutions and infrastructure crumble, no doubt the universities supporting the social sciences and their computer networks storing the deluge of words and data will be among the first things to fall. Indeed, this might account for the propensity of scholarship to rely on optimistic projections; professors and researchers have a strong interest in the viability of current institutions in the face of climate change. Contemplating a world of Deep Adaptation would require imagining a planet on which tenured faculty positions and comfortable offices no longer exist. Of course, the same applies to chief executive officers in offices atop tall buildings in London and Tokyo, and officials with prominent roles in UN offices in New York and Geneva. No wonder few wish to understand adaptation in such terms. With the social practices and physical manifestations of today's institutions swept away, people will not be looking for more research on how to adapt to climate change. Imposed adaptation will be happening, perhaps in a most dramatic way unlikely to be congenial to established careers and the organizations that enable them.

That said, the scholarly enterprise will continue for some time to come, and it is useful to note where it might go next in investigations of adaptation to climate change. It is likely that the increasing attention paid to transformation will only become more intense, as the evidence of harm to societies around the world makes the daily headlines. With that in mind, scholars working on adaptation should take note of large bodies of work on system transformation. Articles appearing in peer-reviewed journals often call for more research on how systems change and the requisites for transformation as if that has not been the main concern of Marxists, socialists, and historians for many decades. Perhaps this book's extensive discussion of capitalism as a hard limit to equitable, effective adaptation to climate change will help to widen the terms of the discussion. In any case, it is not acceptable for scholars to suggest the need for more research on social transformation while neglecting to mention past and current outpourings of writing on that very topic. Scholars need not agree that, say, a class analysis or a world-systems analysis has it right, but they should at least acknowledge and engage with what has been said.

At the same time, those working in radical traditions that do grapple with how to bring about social transformation could devote more attention to what kind of society they expect to follow, and how it will perform better at the nuts and bolts of adapting infrastructure, economy, technology, and social systems to a warmer world. It is not enough to refer to general attributes of a postcapitalist society when a specific challenge will be on the agenda: how to adapt to the climate disaster that neoliberal capitalism has left in its wake. Advocates of ecosocialism and ecological civilization would do well to propose systems of governance and production, beyond local economic democracy and communities, to replace those now dominant. At the global level, the ecosocialist perspective cannot be only about replacing the current managers of the means of production with a new set of managers. Transformation of the social and political structure of society, to be ecosocialist, must also mean transformation of the means of production. If this entails simply scrapping significant elements of current infrastructure and industrial capacity, such as the immense investment in fossil fuel production and consumption, so be it. How to bring about that comprehensive transformation of the physical and the social aspects of the mode of production demands further investigation, giving due regard to work already done on this vast topic.

Judging which is the most useful direction for adaptation research to take depends on what one expects regarding temperature levels, both transient and equilibrium, and what ecological and social impacts one anticipates to follow from a given temperature rise. If UNFCCC negotiations and national policies succeed in holding the rise below the internationally defined danger zone, then identifying the barriers and opportunities for incremental adaptation would be apt. If, as expected, the global average temperature reaches as much as 3.0°C by 2100 and stabilizes, then research should devote more attention to transformational adaptation, in particular on how to bring about the sectoral transitions that the IPCC and scholars consider necessary to achieve climate-resilient development. But if temperatures go well into the danger zone, perhaps reaching 4.0°C by 2100 and rising thereafter, then, if climate change research is possible at all, it will have to focus on how to transform social relations to meet an unprecedented challenge.[59] In light of the unsatisfactory results of global mitigation policy to date, and with little reason to imagine that reality to change enough to matter, transformational change should be the research priority now.

Notes

1. Richard A Slaughter, "Evaluating 'Overshoot and Collapse' Futures," *World Future Review* 2, no. 4 (2010): 5–18.

2. Bill McKibben, *Eaarth: Making Life on a Tough New Planet* (New York: Times Books, 2010), p. 14.

3. Bill McKibben, *Falter: Has the Human Game Begun to Play Itself Out?* (New York: Henry Holt, 2019), Kindle loc. 326.

4. John Michael Greer, *Dark Age America: Climate Change, Cultural Collapse, and the Hard Future Ahead* (Gabriola Island, British Columbia, Canada: New Society, 2016), p. 30.

5. James Lovelock, *The Revenge of Gaia* (New York: Basic Books, 2006), Kindle loc. 128.

6. Anthony J. McMichael and Keith B. Dear, "Climate Change: Heat, Health, and Longer Horizons," *Proceedings of the National Academy of Sciences* 107, no. 21 (2010): 9483–9484; Steven C. Sherwood and Matthew Huber, "An Adaptability Limit to Climate Change Due to Heat Stress," *Proceedings of the National Academy of Sciences* 107, no. 21 (2010): 9552–9555.

7. James Hansen, *Storms of My Grandchildren: The Truth About the Coming Climate Catastrophe and Our Last Chance to Save Humanity* (New York: Bloomsbury Press, 2009).

8. For a widely cited example, see Will Steffen, Johan Rockstram, Katherine Richardson, Timothy M. Lenton, Carl Folke, Dinana Liverman, and Colin P. Summerhayes et al., "Trajectories of the Earth System in the Anthropocene," *Proceedings of the National Academy of Sciences* 115, no. 3 (2018): 8252–8259. For another important statement, see William J. Ripple, Christopher Wolf, Thomas M. Newsome, Phoebe Barnard, and William R. Moomaw, "World Scientists' Warning of a Climate Emergency," *BioScience* 70 no. 1 (2020): 8–12.

9. Rupert Read, "This Civilization Is Finished: Time to Build an Ecological Civilization," *Ecological Citizen* 3, no. 2 (2020): 157–162.

10. Rupert Read, "This Civilization Is Finished: Time to Build an Ecological Civilization," *Ecological Citizen* 3, no. 2 (2020), p. 157, emphasis added.

11. Rupert Read, "This Civilization Is Finished: Time to Build an Ecological Civilization," *Ecological Citizen* 3, no. 2 (2020): 160–161.

12. See, for example, Annalee Newitz, *Scatter, Adapt, and Remember: How Humans Will Survive a Mass Extinction* (New York: Doubleday, 2013).

13. Rupert Read, "This Civilization Is Finished: Time to Build an Ecological Civilization," *Ecological Citizen* 3, no. 2 (2020), p. 159.

14. Jem Bendell, "Deep Adaptation: A Map for Navigating Climate Tragedy," Occasional Paper No. 2 (Carlisle: Institute for Leadership and Sustainability, 2020), p. 13.

15. Jem Bendell, *Breaking Together: A Freedom-Loving Response to Collapse* (Bristol: Good Works, 2023).

16. Raymond Zhong, "Warming Could Push the Atlantic Past a 'Tipping Point' This Century," *New York Times,* July 25, 2023.

17. Catrin Einhorn, "A Desperate Push to Save Florida's Coral: Get It Out of the Sea," *New York Times,* July 31, 2023.

18. Javier Blas, "The Harsh Truth: We're Using More Oil than Ever," *Bloomberg,* July 26, 2023.

19. International Energy Agency, "Oil Market Report—July 2023," 2023, https://www.iea.org.

20. Jem Bendell, "Deep Adaptation: A Map for Navigating Climate Tragedy," Occasional Paper No. 2 (Carlisle: Institute for Leadership and Sustainability), p. 6. Bendell identifies psychological factors and other reasons people refuse to accept the implications of an honest appraisal of climate science.

21. Rupert Read, "Riding Two Horses: The Future of Politics and Activism, as We Face Potential Eco-Driven Societal Collapse," in *Deep Adaptation: Navigating the Realities of Climate Chaos,* edited by Jem Bendell and Rupert Read (Cambridge, England: Polity Press, 2021), pp. 287–288.

22. John Foster, "Hope After Sustainability—Tragedy and Transformation," *Global Discourse* 7, no. 1 (2017): 1–9.

23. Jem Bendell and Rupert Read, "What Next, Now that the Limits Have Been Breached?" in *Deep Adaptation: Navigating the Realities of Climate Chaos,* edited by Jem Bendell and Rupert Read (Cambridge, England: Polity Press, 2021), p. 25.

24. Pablo Servigne, Raphael Stevens, Gauthier Chapelle, and Daniel Rodary, "Reasons for Anticipating Societal Collapse," in *Deep Adaptation: Navigating the Realities of Climate Chaos,* edited by Jem Bendell and Rupert Read (Cambridge, England: Polity Press, 2021), p. 108.

25. Rupert Read, "This Civilization Is Finished: Time to Build an Ecological Civilization," *Ecological Citizen* 3, no. 2 (2020), p. 160.

26. Eddie Yuen, "The Politics of Failure Have Failed: The Environmental Movement and Catastrophism," in *Catastrophism, the Apocalyptic Politics of Collapse*

and Rebirth, edited by Sasha Lilley, David McNally, Eddie Yuen, and James Davis (Toronto, Ontario, Canada: Between the Lines Books, 2012), p. 33.

27. Sean Kelly and Joanna Macy, "The Great Turning: Reconnecting Through Collapse," in *Deep Adaptation: Navigating the Realities of Climate Chaos,* edited by Jem Bendell and Rupert Read (Cambridge, England: Polity Press, 2021), p. 227; see also Robin Globus Veldman, "Narrating the Environmental Apocalypse: How Imagining the End Facilitates Moral Reasoning Among Environmental Activists," *Ethics and the Environment* 17, no. 1 (2012): 1–23.

28. Chris Methmann and Delf Rothe, "Politics for the Day After Tomorrow: The Logic of Apocalypse in Global Climate Politics," *Security Dialogue* 43, no. 4 (2012): 323–344.

29. Jem Bendell, "Deep Adaptation: A Map for Navigating Climate Tragedy," Occasional Paper No. 2 (Carlisle: Institute for Leadership and Sustainability, 2020), p. 22.

30. Matthew Slater and Skeena Rathor, "Relocalization as Deep Adaptation," in *Deep Adaptation: Navigating the Realities of Climate Chaos,* edited by Jem Bendell and Rupert Read (Cambridge, England: Polity Press, 2021): 299–324.

31. Matthew Slater and Skeena Rathor, "Relocalization as Deep Adaptation," in *Deep Adaptation: Navigating the Realities of Climate Chaos*, edited by Jem Bendell and Rupert Read (Cambridge: Polity Press, 2021), p. 317.

32. Jonathan Chait, "Why the Worst Governments in America Are Local Governments," *New York Magazine,* September 2014.

33. John Michael Greer, *Dark Age America: Climate Change, Cultural Collapse, and the Hard Future Ahead* (Gabriola Island, British Columbia, Canada: New Society, 2016), p. 9.

34. Katharine Hayhoe, *Saving Us: A Climate Scientist's Call for Hope and Healing in a Divided World* (New York: One Signal, 2021), p. 115.

35. Paige Bennett, "World Passes Highest Global Average Temperature Recorded," EcoWatch, July 5, 2023, https://www.ecowatch.com; WMO (World Meteorological Organization), "Global Temperatures Set to Reach New Records in Next Five Years," *WMO News,* May 17, 2023.

36. Michael Klare, "Collapse 2.0: What a 2005 Bestseller Tells Us About Climate Change and Human Survival," *Tomdispatch,* August 17, 2023.

37. Sybille Bauriedl and Detlef Muhler-Mann, "The Politics in Critical Adaptation Research," in *A Critical Approach to Climate Change Adaptation,* edited by Silja Klepp and Libertad Chavez-Rodriguez (London: Routledge, 2018), p. 283.

38. John Barkdull and Paul G. Harris, "Climate-Induced Conflict or Hospice Earth: The Increasing Importance of Eco-Socialism," *Global Change, Peace and Security* 27, no. 2 (2015): 237–243.

39. Sven Lütticken, "Unnatural History," *New Left Review* 45 (2007), p. 116.

40. Roger Pielke Jr., "The Political Agenda of the IPCC," *The Honest Broker,* May 15, 2023.

41. Robert Costanza, Gar Alperovitz, Herman Daly, Joshua Farley, Carol Franco, Tim Jackson, Ida Kubiszewski, Juliet Schor, and Peter Victor, *Building a Sustainable and Desirable Economy-in-Society-in-Nature* (Canberra: Australian National University Press, 2013), p. 73.

42. Simon Black, Ian Perry, and Nate Vernon, "Fossil Fuel Subsidies Surged to Record $7 Trillion," *IMF Blog,* August 24, 2023, https://www.imf.org.

43. Jonathan Pickering, Thomas Hickmann, Karin Backstrand, Agni Kalfagianni, Michael Bloomfield, Aysem Mert, Hedda Ransan-Cooper, and Alex Y. Lo, "Democratizing Sustainability Transformations: Assessing the Transformative

Potential of Democratic Practices in Environmental Governance," *Earth System Governance* 11, no. 100131 (2022), p. 10.

44. Elinor Ostrom, *Governing the Commons: The Evolution of Institutions for Collective Action* (Cambridge: Cambridge University Press, 1990).

45. Detlef Sprinz and Tapani Vaahtoranta, "The Interest-Based Explanation of International Environmental Policy," *International Organization* 48, no. 1 (1994): 77–105; John Barkdull, "Nixon and the Marine Environment," *Presidential Studies Quarterly* 27, no. 3 (1998): 587–605.

46. Department of Defense, *Climate Change Adaptation and Resilience,* DOD Directive 4715.21 (2016).

47. John Barkdull, "Responding to Global Warming: Adaptation and Transformational Change," *Journal of the Texas Tech Ethics Center* 3, no. 1 (2019): 77–87.

48. Michael Aklin and Johannes Urpelainen, "The Global Spread of Environmental Ministries: Domestic-International Interactions," *International Studies Quarterly* 58, no. 4 (2014): 764–780.

49. Istvan Meszaros, "The Only Viable Economy," *Monthly Review* 58, no. 11 (2007), pp. 32–33, emphasis in original.

50 Istvan Meszaros, "The Only Viable Economy," *Monthly Review* 58, no. 11 (2007), p. 16.

51. Jason W. Moore, "Ecology, Capital, and the Nature of Our Times: Accumulation & Crisis in the Capitalist World-Ecology," *Journal of World-Systems Research* 17, no. 1 (2011): 107–46.

52. Jason W. Moore, *Capitalism in the Web of Life: Ecology and the Accumulation of Capital* (London: Verso, 2015).

53. Matthew T. Huber, *Climate Change as Class War: Building Socialism on a Warming Planet* (London: Verso, 2022), p. 54.

54. Carl Cassegard, "Activism Without Hope? Four Varieties of Postapocalyptic Environmentalism," *Environmental Politics* 33, no. 3 (2023): 444–464, DOI 10.1080/09644016.2023.2226022; David Chandler, "The Death of Hope? Affirmation in the Anthropocene," *Globalizations* 16, no. 5 (2019): 695–706; Mathias Thaler, "Eco-Miserabilism and Radical Hope: On the Utopian Vision of Post-Apocalyptic Environmentalism," *American Political Science Review* 188, no. 1 (2023): 318–331; Byron Williston, "Climate Change and Radical Hope," *Ethics and the Environment* 17, no. 2 (2012): 166–186.

55. Dan Hurley, "What's the Best Weapon Against Climate Change? Hope," *Newsweek,* August 14, 2023.

56. Marco D'Eramo, "Darning the Planet," *New Left Review Sidecar,* August 8, 2023.

57. Mandy M. Gunasekara, "Environmental Protection Agency," in *Mandate for Leadership: The Conservative Promise,* edited by Paul Dans and Steven Groves (Washington, DC: The Heritage Foundation, 2023), p. 419.

58. This is still not true for political science. Nearly a decade after Debra Javeline's call for political scientists to investigate adaptation, which she deemed the most important topic not being studied, the journal in which she published has carried only a few articles that touch on adaptation and no sustained discussion of the kind Javeline recommended. Debra Javeline, "The Most Important Topic Political Scientists Are Not Studying: Adapting to Climate Change," *Perspectives on Politics* 12, no. 2 (2014): 420–434.

59. John Barkdull and Paul G. Harris, "Climate Catastrophe and Transformationalism," *Seton Hall Journal of Diplomacy and International Relations* 16, no. 1 (2015): 119–130.

Acronyms

AC	Adaptation Committee
AF	Adaptation Fund
AFOLU	agriculture, forestry, and other land use
AGR	Adaptation Gap Report
APA	Ad Hoc Working Group for the Paris Agreement
AR	Assessment Report
AWG-LCA	Ad Hoc Working Group on Long-Term Cooperative Action
CAF	Cancun Adaptation Framework
CBDR	common but differentiated responsibilities
COP	Conference of the Parties to the UN Framework Convention on Climate Change
CO_2	carbon dioxide
CRD	climate-resilient development
EbA	ecosystem-based adaptation
EC	ecological civilization
EMT	ecological modernization theory
EU	European Union
GCF	Global Climate Fund
GDP	gross domestic product
GEF	Global Environmental Fund
GGA	Global Goal on Adaptation
GHG	greenhouse gasses
GMO	genetically modified organism
GNP	gross national product
G-77	Group of 77
ICSU	International Council for Science

IEA	International Energy Agency
IISD	International Institute for Sustainable Development
INC	Intergovernmental Negotiating Committee
INDCs	intended nationally determined contributions
IPCC	Intergovernmental Panel on Climate Change
IR	international relations
IUCN	International Union for Conservation of Nature
L&D	loss and damage
LDCs	least developed countries
LDCF	Least Developed Countries Fund
NAP	National Adaptation Plan
NAPAs	National Adaptation Programmes of Action
NDCs	nationally determined contributions
NGOs	nongovernmental organizations
NWP	Nairobi Work Programme
PAWP	Paris Agreement Work Programme
SBI	Subsidiary Body on Implementation
SBSTA	Subsidiary Body for Scientific and Technological Advice
SCALA	Scaling Up Climate Ambition on Land Use and Agriculture
SDGs	Sustainable Development Goals
SES	socioecological system
UKCIP	UK Climate Impacts Programme
UNCED	UN Conference on Environment and Development
UNDP	UN Development Programme
UNEP	UN Environment Programme
UNFCCC	UN Framework Convention on Climate Change
WCP	World Climate Programme
WCRP	World Climate Research Programme
WIM	Warsaw International Mechanism
WMO	World Meteorological Organization
WSDEs	workers self-directed enterprises
WWF	World Wildlife Fund
XR	Extinction Rebellion

Bibliography

Adger, W. Neil. "Institutional Adaptation to Environmental Risk Under the Transition in Vietnam." *Annals of the Association of American Geographers* 90, no. 4 (2000): 738–758.

Adger, W. Neil. "Vulnerability." *Global Environmental Change* 16, no. 3 (2006): 268–281.

Ahearn, Gerard. "Towards an Ecological Civilization: A Gramscian Strategy for a New Political Subject." *Cosmos and History* 9, no. 1 (2013): 317–326.

Ahmed, Sara, and Elizabeth Fajber. "Engendering Adaptation to Climate Variability in Gujarat, India." *Gender and Development* 17, no. 1 (2009): 33–50.

Aklin, Michael, and Johannes Urpelainen. "The Global Spread of Environmental Ministries: Domestic-International Interactions." *International Studies Quarterly* 58, no. 4 (2014): 764–780.

Albert, Michael J. "Capitalism and Earth System Governance: An Ecological Marxist Approach." *Global Environmental Governance* 20, no. 2 (2020): 37–56.

Albert, Michael J. "Ecosocialism for Realists: Transitions, Trade-Offs, and Authoritarian Dangers." *Capitalism Nature Socialism* 34, no. 1 (2023): 11–30.

Allan, Bentley B. and Jonas O. Meckling. "Creative Learning and Policy Ideas: The Global Rise of Green Growth." *Perspectives on Politics* 21, no. 2 (2022): 443–461.

Anderies, John, Carl Folke, and Elinor Ostrom. "Aligning Key Concepts for Global Change Policy: Robustness, Resilience, and Sustainability." *Ecology and Society* 18, no. 2 (2013): 8.

Anderson, Kevin. "The Hidden Agenda: How Veiled Techno-Utopias Shore Up the Paris Agreement." January 6, 2016. kevinanderson.info/blog.

Anderson, Kevin, John F. Broderick, and Isak Stoddard. "A Factor of Two: How the Mitigation Plans of 'Climate Progressive' Nations Fall Far Short of Paris-Compliant Pathways." *Climate Policy* 20, no. 10 (2020): 1290–1304.

Angus, Ian. *Facing the Anthropocene: Fossil Capitalism and the Crisis of the Earth System*, Kindle ed. New York: Monthly Review Press, 2016.

Antwi-Agyei, Philip, Andrew J. Dougill, Lindsay C. Stringer and Samuel Nii Ardey Codjoe. "Adaptation Opportunities and Maladaptive Outcomes in Climate Vulnerability Hotspots of Northern Ghana." *Climate Risk Management* 7, no. 4 (2018): 83–93.

Aufenvenne, Philipp, Heike Egner, and Kirsten von Elverfeldt. "On Climate Change Research, the Crisis of Science and Second-Order Science." *Constructivist Foundations* 10, no. 1 (2014): 120–139.

Ayers, Jessica, and David Dodman. "Climate Change Adaptation and Development, pt. I: The State of the Debate." *Progress in Development Studies* 10, no. 2 (2010): 161–168.

Baran, Paul A., and Paul M. Sweezy. *Monopoly Capital: An Essay on the American Economic and Social Order*. New York: Monthly Review Press, 1966.

Bardsley, Douglas K., and Graeme J. Hugo. "Migration and Climate Change: Examining Thresholds of Change to Guide Effective Adaptation Decision-Making." *Population and Environment* 32, nos. 2–3 (2010): 238–262.

Barkdull, John. "Nixon and the Marine Environment." *Presidential Studies Quarterly* 27, no. 3 (1998): 587–605.

Barkdull, John. "Responding to Global Warming: Adaptation and Transformational Change." *Journal of the Texas Tech Ethics Center* 3, no. 1 (2019): 77–87.

Barkdull, John. "The Ethics of Sustainability." *Journal of the Texas Tech Ethics Center* 4, no. 2 (2020).

Barkdull, John, and Paul G. Harris. "The Land Ethic: A New Philosophy for International Relations." *Ethics and International Affairs* 12, no. 1 (1998): 159–177.

Barkdull, John, and Paul G. Harris. "Climate Catastrophe and Transformationalism." *Seton Hall Journal of Diplomacy and International Relations* 16, no. 1 (2015): 119–130.

Barkdull, John, and Paul G. Harris. "Climate-Induced Conflict or Hospice Earth: The Increasing Importance of Eco-Socialism." *Global Change, Peace and Security* 27, no. 2 (2015): 237–243.

Barkdull, John, and Paul G. Harris. "Political Science and Severe Climate Change: A Case for Transformational Research into Adaptation." *St. Antony's International Review* 13, no. 1 (2017): 120–140.

Barkdull, John, and Paul G. Harris. "Emerging Responses to Global Climate Change: Ecosystem-Based Adaptation." *Global Change, Peace and Security* 31, no. 2 (2019): 19–37.

Barrett, Sam. "Local Level Climate Justice? Adaptation Finance and Vulnerability Reduction." *Global Environmental Change* 23, no. 6 (2013): 1819–1829.

Barry, John. "Green Republicanism and a 'Just Transition' from the Tyranny of Economic Growth." *Critical Review of International Social and Political Philosophy* 24, no. 5 (2019): 725–742.

Bauriedl, Sybille, and Detlef Muhler-Mann. "The Politics in Critical Adaptation Research." In *A Critical Approach to Climate Change Adaptation,* edited by Silja Klepp and Libertad Chavez-Rodriguez, 275–287. London: Routledge, 2018.

Bazilian, Morgan, and Dolf Gielen. "5 Years After Paris: How Countries' Climate Policies Match Up to Their Promises, and Who's Aiming for Net Zero Emissions." *The Conversation,* December 10, 2020.

Bendell, Jem. "Deep Adaptation: A Map for Navigating Climate Tragedy." Occasional Paper No. 2. Carlisle: Institute for Leadership and Sustainability, 2020.

Bendell, Jem. *Breaking Together: A Freedom-Loving Response to Collapse*. Bristol: Good Works, 2023.

Bendell, Jem, and Rupert Read, eds. *Deep Adaptation: Navigating the Realities of Climate Chaos*. Cambridge, England: Polity Press, 2021.

Bendell, Jem, and Rupert Read. "What Next, Now that the Limits Have Been Breached?" In *Deep Adaptation: Navigating the Realities of Climate Chaos,* edited by Jem Bendell and Rupert Read, 20-30. Cambridge, England: Polity Press, 2021.

Benson, Melinda Harm, and Robin Kundis Craig. *The End of Sustainability: Resilience and the Future of Environmental Governance in the Anthropocene*. Lawrence: University of Kansas Press, 2017.

Benton, Ted. "Ecology, Socialism and the Mastery of Nature: A Reply to Reiner Grundmann," *New Left Review* 194 (1992).

Bhaskar, Roy, Cheryl Frank, Karl Georg Høyer, Petter Næss, and Jenneth Parker, eds. *Interdisciplinarity and Climate Change: Transforming Knowledge and Practice for Our Global Future*. London: Routledge, 2010.

Biagini, Bonizella, Rosina Bierbaum, Missy Stults, Saliha Dobardzic, and Shannon M. McNeeley. "A Typology of Adaptation Actions: A Global Look at Climate Adaptation Actions Financed Through the Global Environment Facility." *Global Environmental Change* 25 (2014): 97–108.

Biermann, Frank, Kenneth Abbott, Steinar Andresen, Karen Bäckstrand, Steven Bernstein, Michelle M. Betsill, and Harriet Bulkeley et al. "Navigating the Anthropocene: Improving Earth System Governance." *Science* 335, no. 6074 (2012): 1306–1307.

Biesbroek, G. Robert, , Judith E. M. Klostermann, Catrien J. A. M. Termeer, and Pavel Kabat. "On the Nature of Barriers to Climate Change Adaptation." *Regional Environmental Change* 13, no. 5 (2013): 1119–1129.

Blackwater, Bill. "The Denialism of Progressive Environmentalists." *Monthly Review* 64, no. 2 (2012): 10–21.

Boehm, Sophie, Katie Lebling, Kelly Levin, Hanna Fekete, Joel Jaeger, Anna Nilsson, Ryan Wilson, Andreas Geiges, and Clea Schumer. *State of Climate Action 2021: Systems Transformations Required to Limit Global Warming to 1.5°C*. Washington, DC: World Resources Institute, 2021.

Boggs, Carl. "The Grand Illusion." *Counterpunch,* November 29, 2019.

Borowy, Iris. "Sustainability and Redistribution." In *What Next for Sustainable Development: Our Common Future at Thirty,* edited by James Meadowcroft, David Banister, Erling Holden, Oluf Langhelle, Kristin Linnerud, and Geoffrey Gilpin, 120-137. Cheltenham: Edward Elgar, 2019.

Boulton, Chris A., Timothy M. Lenton, and Niklas Boers. "Pronounced Loss of Amazon Rainforest Resilience Since the Early 2000s." *Nature Climate Change* 12, no. 3 (2022): 271–278.

Brooks, Nick, and W. Neil Adger. "Assessing and Enhancing Adaptive Capacity." In *Adaptation Policy Frameworks for Climate Change: Developing Strategies, Policies and Measures,* edited by Bo Lim and Erika Spanger-Siegfried, 165–181. Cambridge, England: Cambridge University Press, 2004.

Brown, Katrina. "Sustainable Adaptation: An Oxymoron?" *Climate and Development* 3, no. 1 (2011): 21–31.

Brunnengräber, Achim. "The Political Economy of the Kyoto Protocol." In *Socialist Register 2007: Coming to Terms with Nature,* edited by Leo Panitch and Colin Leys, 213–230. London: Merlin Press, 2006.

Brysse, Keynyn, Naomi Oreske, Jessica O'Reilly, and Michael Oppenheimer. "Climate Change Prediction: Erring on the Side of Least Drama?" *Global Environmental Change* 23, no. 1 (2013): 327–337.

Burkett, Paul. "An Eco-Revolutionary Tipping Point?" *Monthly Review* 69, no. 1 (2017): 1–19.

Büscher, Bram, and Robert Fletcher. *The Conservation Revolution: Radical Ideas for Saving Nature Beyond the Anthropocene*. Kindle ed. London: Verso, 2020.

Campos, Inis S., Filipe M. Alves, Joao Dinis, Monica Truninger, Andre Vizinho, and Gil Penha-Lopes. "Climate Adaptation, Transitions, and Socially Innovative Action-Research Approaches." *Ecology and Society* 21, no. 1 (2016): 13.

Cannavó, Peter F. "In the Wake of Katrina: Climate Change and the Coming Crisis of Displacement." In *Political Theory and Global Climate Change,* edited by Steve Vanderheiden, 177–200. Cambridge: MIT Press, 2008.

Carpenter, Stephen R., and William A. Brock. "Adaptive Capacity and Traps." *Ecology and Society* 13, no. 2 (2008): 40.

Carr, Edward R. "Properties and Projects: Reconciling Resilience and Transformation for Adaptation and Development." *World Development* 122 (2019): 70–84.

Cassegard, Carl. "Activism Without Hope? Four Varieties of Postapocalyptic Environmentalism." *Environmental Politics* 33, no. 3 (2023): 444-464. DOI 10.1080/09644016.2023.2226022.

Chaffin, Brian C., Hannah Gosnell, and Barbara A. Cosens. "A Decade of Adaptive Governance Scholarship: Synthesis and Future Directions." *Ecology and Society* 19, no. 3 (2014): 56.

Chandler, David. "The Death of Hope? Affirmation in the Anthropocene." *Globalizations* 16, no. 5 (2019): 695–706.

Chen, Ying. "Degrowth—What's in a Name? Assessing Degrowth's Political Implications." *Monthly Review* 75, no. 3 (2023): 160–170.

Ciplet, David, J. Timmons Roberts, and Mizan R. Khan. "The Politics of International Climate Adaptation Funding: Justice and Divisions in the Greenhouse." *Global Environmental Politics* 13, no. 1 (2013): 49–68.

Colloff, Matthew J., Berta Martín-López, Sandra Lavorel, Bruno Locatelli, Russell Gorddard, Pierre-Yves Longaretti, and Gretchen Walters et al. "An Integrative Research Framework for Enabling Transformative Adaptation." *Environmental Science and Policy* 68 (2017): 87–96.

Committee on Understanding and Monitoring Abrupt Climate Change and Its Impacts. *Abrupt Impacts of Climate Change: Anticipating Surprises*. Washington, DC: National Academies Press, 2013.

Costanza, Robert, Gar Alperovitz, Herman Daly, Joshua Farley, Carol Franco, Tim Jackson, Ida Kubiszewski, Juliet Schor, and Peter Victor. *Building a Sustainable and Desirable Economy-in-Society-in-Nature*. Canberra: Australian National University Press, 2013.

Costanza, Robert, Joshua Farley, and Ida Kubiszewski. "Adapting Institutions for Life in a Full World." In *State of the World Atlas 2010,* edited by E. Assadourian, 85–90. Washington, DC: Worldwatch, 2010.

Cox, Robert W. "Social Forces, States and World Orders: Beyond International Relations Theory." *Millennium: Journal of International Studies* 10, no. 2 (1981): 126–155.

Croeser, Eve. *Ecosocialism and Climate Justice: An Ecological Neo-Gramscian Analysis.* Abingdon: Routledge, 2021.

Croeser, Eve. "From Climate Change to Climate Action to Climate Justice: An Ecological Neo-Gramscian Analysis of Ecosocialism's Potential." Doctoral dissertation, University of Tasmania, 2017.

Dale, Lisa. *Climate Change Adaptation: An Earth Institute Sustainability Primer.* New York: Columbia University Press, 2022.

Daly, Herman. "Toward Some Operational Principles of Sustainable Development." *Ecological Economics* 2, no. 1 (1990): 1–6.

Daly, Herman. *Steady-State Economics: Second Edition with New Essays*. Washington, DC: Island Press, 1991.

Davis, Mike. "Who Will Build the Ark?" *New Left Review* 61 (2010): 29–46.

D'Eramo, Marco. "Darning the Planet." *New Left Review Sidecar,* August 8, 2023.

De Lucia, Vito. "The Climate Justice Movement and the Hegemonic Discourse of Technology." In *Routledge Handbook of the Climate Change Movement,* edited by Matthias Dietz and Heiko Garrelts, 66–83. London: Routledge, 2013.

de Zoysa, Kiyomi, Tamara Coger, and Nisha Krishnan. "Can the Global Goal on Adaptation Be Locally Led?" Washington, DC: World Resources Institute, July 22, 2022.

Dimitrov, Radoslav S. "Inside UN Climate Change Negotiations: The Copenhagen Conference." *Review of Policy Research* 27, no. 6 (2010): 795–821.

Dow, Kirstin, Frans Berkhout, Benjamin L. Preston, Richard J. T. Klein, Guy Midgley, and M. Rebecca Shaw. "Limits to Adaptation." *Nature Climate Change* 3, no. 4 (2013): 305–307.

Ehrlich, Paul R., and Anne H. Ehrlich. "Can a Collapse of Global Civilization Be Avoided?" *Proceedings of the Royal Society B* 280, no. 20122845 (2013): 1–9.

Ehrlich, Paul R., and Dennis C. Pirages. "Political Science in a New Era." *World Future Review* 4, no. 3 (2012): 16–22.

Ellicott, Val. "Climate Change: Will World Governments Act in Time?" *CQ Researcher,* October 25, 2021.

Engle, Nathan L. "Adaptive Capacity and Its Assessment." *Global Environmental Change* 21, no. 2 (2011): 647–656. Enroth, Henrik. "Declarations of Dependence: On the Constitution of the Anthropocene." *Theory, Culture and Society* 38, nos. 7–8 (2020): 1–22.

Eriksen, Siri H., Katrina Brown, and P. Mick Kelly. "The Dynamics of Vulnerability: Locating Coping Strategies in Kenya and Tanzania." *Geographical Journal* 171, no. 4 (2005): 287–305.

Eriksen, Siri H., Andrea J. Nightingale, and Hallie Eakin. "Reframing Adaptation: The Political Nature of Climate Change Adaptation." *Global Environmental Change* 35 (2015): 523–533.

Fankhauser, Samuel, and Guido Schmidt-Traub. "From Adaptation to Climate-Resilient Development: The Costs of Climate-Proofing the Millennium Development Goals in Africa." *Climate and Development* 3, no. 2 (2011): 94–113.

Fannelli, Carlo. "Climate Change: 'the Greatest Challenge of Our Time.'" *Alternate Routes* 14 (2014): 15–31.

Farber, Daniel A. "Adapting to Climate Change: Who Should Pay." *Journal of Land Use and Environmental Law* 23, no. 1 (2007): 1–37.

Fazey, Ioan, Peter Moug, Simon Allen, Kate Beckmann, David Blackwood, Mike Bonaventura, and Kathryn Burnett et al., "Transformation in a Changing Climate: A Research Agenda." *Climate and Development* 10, no. 3 (2018): 197–217.

Fedorov, Evgenii K. "Climatic Change and Human Strategy." In *Proceedings of the World Climate Conference: A Conference of Experts on Climate and Mankind,* 15–26. Geneva: Secretariat of the World Meteorological Organization, 1979.

Feola, Giuseppe. "Societal Transformation in Response to Global Environmental Change: A Review of Emerging Concepts." *Ambio* 44, no. 5 (2015): 376–390.

Fischer, Dana R., and William R. Freudenburg. "Ecological Modernization and Its Critics: Assessing the Past and Looking Toward the Future." *Society and Natural Resources* 14 (2001): 701–709.

Fischer, Frank. "Knowledge Politics and Post-Truth in Climate Denial: On the Social Construction of Alternative Facts." *Critical Policy Studies* 13, no. 2 (2019): 133–152.

Fisher, Dana R. and William R. Freudenburg. "Ecological Modernization and Its Critics: Assessing the Past and Looking Toward the Future." *Society and Natural Resources* 14, no. 8 (2001): 701–709.
Folke, Carl. "How Resilient Are Ecosystems to Global Environmental Change?" *Sustainability Science* 5 (2010): 151–154.
Folke, Carl, Stephen R. Carpenter, Brian Walker, Martin Scheffer, Terry Chapin, and Johan Rockstrom. "Resilience Thinking: Integrating Resilience, Adaptability and Transformability." *Ecology and Society* 15, no. 4 (2010): 20.
Ford, James. "Emerging Trends in Climate Change Policy: The Role of Adaptation." *International Public Policy Review* 3, no. 2 (2007): 5–16.
Foster, John. "Hope After Sustainability—Tragedy and Transformation." *Global Discourse* 7, no. 1 (2017): 1–9.
Foster, John Bellamy. "Capitalism and the Curse of Energy Efficiency: The Return of the Jevons Paradox." *Monthly Review* 62, no. 6 (2010): 1–12.
Foster, John Bellamy. "Capitalism and Degrowth: An Impossibility." *Monthly Review* 62, no. 8 (2011): 26–33.
Foster, John Bellamy. "The Planetary Rift and the New Human Exemptionalism: A Political-Economic Critique of Ecological Modernization Theory." *Organization and Environment* 25, no. 3 (2012): 211–237.
Foster, John Bellamy. "The Earth-System Crisis and Ecological Civilization: A Marxian View." *International Critical Thought* 7, no. 4 (2017): 439–458.
Foster, John Bellamy. "The Long Ecological Revolution." *Monthly Review* 69, no. 6 (2017): 1–16.
Foster, John Bellamy. "Making War on the Planet: Geoengineering and Capitalism's Creative Destruction of the Earth." *Monthly Review* 70, no. 4 (2018): 1–10.
Foster, John Bellamy. "Capitalism Has Failed—What Next?" *Monthly Review* 70, no. 9 (2019): 1–24.
Foster, John Bellamy. "Ecological Civilization, Ecological Revolution." *Monthly Review* 74, no. 5 (2022): 1–11.
Foster, John Bellamy. "Planned Degrowth: Ecosocialism and Sustainable Human Development." *Monthly Review* 75, no. 3 (2023): 1–29.
Foster, John Bellamy, Brett Clark, and Richard York. *The Ecological Rift: Capitalism's War on the Earth*. New York: Monthly Review Press, 2010. Chapter 1.
Fraser, Nancy. "Behind Marx's Hidden Abode." *New Left Review* 86 (2014): 55–72.
Fraser, Nancy. "Climates of Capital: For a Trans-Environmental Eco-Socialism." *New Left Review* 127 (2021): 94–127.
Füssel, Hans-Martin "Adaptation Planning for Climate Change: Concepts, Assessment Approaches, and Key Lessons." *Sustainability Science* 2, no. 2 (2007): 265–275.
Füssel, Hans-Martin. "How Inequitable Is the Global Distribution of Responsibility, Capability, and Vulnerability to Climate Change: A Comprehensive Indicator-Based Assessment." *Global Environmental Change* 20, no. 4 (2010): 597–611.
Gare, Arran. *Nihilism Incorporated: European Civilization and Environmental Destruction.* Bungendore, Australia: Eco-Logical Press, 1993.
Gare, Arran. "Barbarity, Civilization and Decadence: Meeting the Challenge of Creating an Ecological Civilization." In *Yearbook of Philosophy in Process,* edited by Michel Weber and Ronny Desmet, 167–189. Louvain-la-Neuve, Belgium: Presses Universitaires de Louvain, 2009.
Gare, Arran. "From 'Sustainable Development' to 'Ecological Civilization': Winning the War for Survival." *Cosmos and History* 13, no. 3 (2017): 130–153.

Gare, Arran. "After Neoliberalism: From Eco-Marxism to Ecological Civilization, Part 2." *Capitalism Nature Socialism* 32, no. 3 (2021): 43–60.

Gates, W. Lawrence. "The Physical Basis of Climate." In *Proceedings of the World Climate Conference: A Conference of Experts on Climate and Mankind,* World Meteorological Organization, 112–131. Geneva: Secretariat of the World Meteorological Organization, 1979.

Gilbert, Chris. "'Where Danger Lies....' The Communal Alternative in Venezuela." *Monthly Review* 75, no. 3 (2023): 72–84.

Gillard, Ross, Andrew Gouldson, Jouni Paavola, and James Van Alstine. "Transformational Responses to Climate Change: Beyond a Systems Perspective of Social Change in Mitigation and Adaptation." *Wiley Interdisciplinary Reviews: Climate Change* 7, no. 2 (2016): 251–265.

Global Commission on the Economy and Climate. *The New Climate Economy.* Washington, DC: World Resources Institute, 2018.

Goklany, Indur M. "Integrated Strategies to Reduce Vulnerability and Advance Adaptation, Mitigation, and Sustainable Development." *Mitigation and Adaptation Strategies for Global Change* 12, no. 5 (2007): 755–786.

Görg, Christoph, Ulrich Brand, Helmut Haberl, Diana Hummel, Thomas Jahn, and Stefan Liehr. "Challenges for Social-Ecological Transformations: Contributions from Social and Political Ecology." *Sustainability* 9, no. 7: 1045 (2017). DOI 10.3390/su9071045.

Granados, Jose A. Tapia. "Inexorable March Toward Utter Climate Disaster?" *Climate Nature Socialism* 29, no. 4 (2018): 21–30.

Gray, C. Boyden, and David B. Rivkin Jr. "A 'No Regrets' Environmental Policy." *Foreign Policy* 83 (1991): 47–65.

Greer, John Michael. *Dark Age America: Climate Change, Cultural Collapse, and the Hard Future Ahead.* Gabriola Island, British Columbia, Canada: New Society, 2016.

Guerrero, Dorothy Grace. "The Limits of Capitalist Solutions to the Climate Crisis." In *The Climate Crisis: South African and Global Democratic Eco-Socialist Alternatives,* edited by Vishwas Satgar, 30–46. Johannesburg: Wits University Press, 2018.

Gunasekara, Mandy M. "Environmental Protection Agency." In *Mandate for Leadership: The Conservative Promise,* edited by Paul Dans and Steven Groves. 417–448.

Haas, Peter M. *Saving the Mediterranean: The Politics of International Environmental Cooperation.* New York: Columbia University Press, 1990.

Hamer, Mick. "Plot to Undermine Global Pollution Controls Revealed." *New Scientist,* January 2, 2002.

Hansen, James. *Storms of My Grandchildren: The Truth About the Coming Climate Catastrophe and Our Last Chance to Save Humanity.* New York: Bloomsbury Press, 2009.

Hansen, James, Pushker Kharecha, Makiko Sato, Valerie Masson-Delmotte, Frank Ackerman, David J. Beerling, and Paul J. Hearty et al. "Assessing 'Dangerous Climate Change': Required Reduction of Carbon Emissions to Protect Young People, Future Generations, and Nature." *PLOS ONE* 8, no. 12 (2013): e81648.

Harmeling, Sven, and Christoph Bals. "Adaptation to Climate Change—Where Do We Go from Bali?" Briefing Paper. Bonn: Germanwatch, 2008.

Harris, Paul G. "Collective Action on Climate Change: The Logic of Regime Failure." *Natural Resources Journal* 47, no. 1 (2007): 195–224.

Harris, Paul G. *What's Wrong with Climate Politics and How to Fix It*. Cambridge, England: Polity, 2013.
Harris, Paul G. "COP26: The Eternally Weak Pulse of Climate Diplomacy, and What Needs to Change." *PLOS Climate* 1, no. 3 (2022): e0000019.
Harris, Paul G., and John Barkdull. "Adapting to Dangerous Climate Change: Implications for Studies of Politics, Policy and Beyond." *Journal of Environmental Studies and Sciences* 6, no. 1 (2016): 451–459.
Harvey, David. *A Brief History of Neoliberalism*. Oxford: Oxford University Press, 2005.
Harvey, Mark. "Climate Emergency: How the Inequality Crisis Is Dynamically Linked to the Sociogenesis of Climate Change." *Global Social Challenges Journal* 1, no. 1 (2022): 115–137.
Hayhoe, Katharine. *Saving Us: A Climate Scientist's Call for Hope and Healing in a Divided World*. New York: One Signal, 2021.
Hermwille, Lukas. "Climate Change as a Transformation Challenge: A New Climate Policy Paradigm?" *GAIA—Ecological Perspectives for Science and Society* 25, no. 1 (2016): 19–22.
Herrick, Charles N. "Science and Climate Policy: A History Lesson." *Issues in Science and Technology* 8, no. 2(1991): 56–57.
Hickel, Jason. "What Does Degrowth Mean? A Few Points of Clarification." *Globalizations* 18, no. 7 (2021): 1105–1111.
Hickel, Jason, and Giorgos Kallis. "Is Green Growth Possible?" *New Political Economy* 25, no. 4 (2020): 469–486.
Hickel, Jason. "The Anti-Colonial Politics of DeGrowth." *Political Geography* 88 (2021): 1024040.
Hölscher, Katharina, Julia M. Wittmayer, and Derek Loorbach. "Transition Versus Transformation: What's the Difference?" *Environmental Innovation and Societal Transitions* 27 (2018): 1–3.
Huber, Matthew T., *Climate Change as Class War: Building Socialism on a Warming Planet*. London: Verso, 2022.
IISD (International Institute for Sustainable Development). "Overview of the UNCED Process." *Earth Summit Bulletin* 2, no. 1 (1992), p. 1.
IISD. "Seventh Conference of the Parties to the UN Framework Convention on Climate Change: Monday, 29 October 2001." *Earth Negotiations Bulletin* 12, no. 179 (2001).
IISD. "UNFCCC COP-7 Highlights: Wednesday, 7 November 2001." *Earth Negotiations Bulletin* 12, no. 187 (2001).
IISD. "UNFCCC COP-7 Highlights: Thursday, 8 November 2001." *Earth Negotiations Bulletin,* 12, no. 188 (2001).
IISD. "Summary of the Tenth Conference of the Parties to the UN Framework Convention on Climate Change, 6–18 December 2004." *Earth Negotiations Bulletin* 12, no. 260 (2004).
IISD. "Summary of the Eleventh Conference of the Parties to the UN Framework Convention on Climate Change and the First Conference of the Parties Serving as the Meeting of the Parties to the Kyoto Protocol: 28 November–10 December 2005." *Earth Negotiations Bulletin* 12, no. 291 (2005).
IISD. "Summary of the Twelfth Conference of the Parties to the UN Framework Convention on Climate Change and the Second Meeting of the Parties to the Kyoto Protocol: 6–17 November 2006." *Earth Negotiations Bulletin* 12, no. 318 (2006).

IISD. "Summary of the Thirteenth Conference of Parties to the UN Framework Convention on Climate Change and Third Meeting of the Parties to the Kyoto Protocol: 3–15 December 2007." *Earth Negotiations Bulletin* 12, no. 354 (2007).

IISD. "Summary of the Copenhagen Climate Change Conference, 7–19 December 2009." *Earth Negotiations Bulletin* 12, no. 459 (2009).

IISD. "Summary of the Cancun Climate Change Conference, 29 November–11 December 2010." *Earth Negotiations Bulletin* 12, no. 498 (2010).

IISD. "Summary of the Durban Climate Change Conference 28 November–11 December 2011." *Earth Negotiations Bulletin* 12, no. 534 (2011).

IISD. "Summary of the Lima Climate Change Conference—1–14 December 2014." *Earth Negotiations Bulletin* 12, no. 619 (2014).

IISD. "Paris Highlights: Monday, 7 December 2015." *Earth Negotiations Bulletin* 12, no. 659 (2015).

IISD. "Summary of the Paris Climate Change Conference: 29 November–13 December 2015." *Earth Negotiations Bulletin* 12, no. 663 (2015).

IISD, "Summary Report, 2–15 December 2018: Katowice Climate Change Conference—December 2018," *Earth Negotiations Bulletin* 12, no. 747 (2018).

IISD. "Summary of the Chile/Madrid Climate Change Conference: 2–15 December 2019." *Earth Negotiations Bulletin* 12, no. 775 (2019).

IISD. "Glasgow Climate Change Conference: 31 October–13 November 2021." *Earth Negotiations Bulletin* 12, no. 793 (2021).

IISD. "Summary of the Sharm el-Sheikh Climate Change Conference: 6–20 November 2022." *Earth Negotiations Bulletin* 12, no. 818 (2022).

IISD. "Summary of the 2023 Dubai Climate Change Conference: 30 November–13 December 2023." *Earth Negotiations Bulletin* 12, no. 842 (2023).

International Energy Agency. *Coal 2022: Analysis and Forecast to 2025.* December 2022. https://iea.blob.core.windows.net/.

Initiative for Climate Action Transparency. *Transformational Change Methodology: Assessing the Transformational Impacts of Policies and Actions.* Edited by Karen H. Olsen and Neelam Singh (Washington, DC: World Resources Institute, 2020).

International Energy Agency. "Global Oil Market—July 2023." 2023. https://www.iea.org.

IPCC (Intergovernmental Panel on Climate Change). *Climate Change: The IPCC response strategies.* Washington, DC: Island Press, 1991.

IPCC. *Climate Change: The 1990 and 1992 IPCC Assessments.* Canada: World Meteorological Organization (WMO) and UN Environment Programme (UNEP), 1992.

IPCC. *Climate Change 1995: Impacts, Adaptations and Mitigation of Climate Change: Scientific-Technical Analyses. Contribution of Working Group II to the Second Assessment Report of the Intergovernmental Panel on Climate Change.* Cambridge: Cambridge University Press, 1995.

IPCC. *IPCC Second Assessment: Climate Change 1995.* Geneva: IPCC, 1995.IPCC. *Climate Change 1995: The Science of Climate Change. Contribution of WGI to the Second Assessment Report of the Intergovernmental Panel on Climate Change.* Cambridge: Cambridge University Press, 1996.

IPCC. *Climate Change 2001: Impacts, Adaptation and Vulnerability. Contribution of Working Group II to the Third Assessment Report of the Intergovernmental Panel on Climate Change.* Cambridge: Cambridge University Press, 2001.

IPCC. *Climate Change 2001: Synthesis Report.* Cambridge: Cambridge University Press, 2001.

IPCC. *Climate Change 2007: Impacts, Adaptation and Vulnerability. Contribution of Working Group II to the Fourth Assessment Report of the Intergovernmental Panel on Climate Change*. Cambridge: Cambridge University Press, 2007.

IPCC. *Climate Change 2007: Synthesis Report. Contribution of Working Groups I, II and III to the Fourth Assessment Report of the Intergovernmental Panel on Climate Change.* Geneva: IPCC, 2007.

IPCC. "Perspectives on Climate Change and Sustainability." In *Climate Change 2007: Impacts, Adaptation and Vulnerability. Contribution of Working Group II to the Fourth Assessment Report of the Intergovernmental Panel on Climate Change.* Cambridge: Cambridge University Press, 2007: 811–841.

IPCC. "Assessment of Adaptation Practices, Options, Constraints and Capacity." In *Climate Change 2007: Impacts, Adaptation and Vulnerability. Contribution of Working Group II to the Fourth Assessment Report of the Intergovernmental Panel on Climate Change.* Cambridge: Cambridge University Press, 2007: 717–743.

IPCC. "Inter-Relationships Between Adaptation and Mitigation." In *Climate Change 2007: Impacts, Adaptation and Vulnerability. Contribution of Working Group II to the Fourth Assessment Report of the Intergovernmental Panel on Climate Change.* Cambridge: Cambridge University Press, 2007: 745–777.

IPCC. "Long-Term Climate Change: Projections, Commitments and Irreversibility." In *Climate Change 2013: The Physical Science Basis. Contribution of Working Group I to the Fifth Assessment Report of the Intergovernmental Panel on Climate Change*. Cambridge: Cambridge University Press, 2013: 1029–1136.

IPCC. "Adaptation Opportunities, Constraints, and Limits." In *Climate Change 2014: Impacts, Adaptation and Vulnerability. Contribution of Working Group II to the Sixth Assessment Report of the Intergovernmental Panel on Climate Change.* Cambridge: Cambridge University Press, 2014: 899–944.

IPCC. *Climate Change 2014: Impacts, Adaptation and Vulnerability. Contribution of Working Group II to the Fifth Assessment Report of the Intergovernmental Panel on Climate Change.* Cambridge: Cambridge University Press, 2014.

IPCC. *Climate Change 2014: Synthesis Report. Contribution of Working Groups I, II and III to the Fifth Assessment Report of the Intergovernmental Panel on Climate Change*. Geneva: IPCC, 2014.

IPCC. *Climate Change 2014: Synthesis Report. Summary for Policymakers*. Geneva: IPCC, 2014.

IPCC. "Climate-Resilient Pathways: Adaptation, Mitigation, and Sustainable Development." In *Climate Change 2014: Impacts, Adaptation, and Vulnerability. Part A: Global and Sectoral Aspects. Contribution of Working Group II to the Fifth Assessment Report of the Intergovernmental Panel on Climate Change,* 1101–1131. Cambridge: Cambridge University Press, 2014.

IPCC. "Foundations for Decision-Making." In *Climate Change 2014: Impacts, Adaptation and Vulnerability. Contribution of Working Group II to the Sixth Assessment Report of the Intergovernmental Panel on Climate Change.* Cambridge: Cambridge University Press, 2014: 195–228.

IPCC. "Assessing Transformation Pathways." In *Climate Change 2014: Mitigation of Climate Change. Contribution of Working Group III to the Fifth Assessment Report of the Intergovernmental Panel on Climate Change*. Cambridge: Cambridge University Press: 413–510.

IPCC. "Summary for Policymakers." In *Climate Change 2021: The Physical Science Basis. Contribution of Working Group I to the Sixth Assessment Report of the Intergovernmental Panel on Climate Change,* 3–31. Cambridge: Cambridge University Press, 2021.

IPCC. *Climate Change 2022: Impacts, Adaptation, and Vulnerability. Contribution of Working Group II to the Sixth Assessment Report of the Intergovernmental Panel on Climate Change.* Cambridge: Cambridge University Press, 2022.

IPCC. "Climate Resilient Development Pathways." In *Climate Change 2022: Impacts, Adaptation and Vulnerability. Contribution of Working Group II to the Sixth Assessment Report of the Intergovernmental Panel on Climate Change,* 2655–2807. Cambridge: Cambridge University Press, 2022.

IPCC. "Decision-Making Options for Managing Risk." In *Climate Change 2022: Impacts, Adaptation and Vulnerability. Contribution of Working Group II to the Sixth Assessment Report of the Intergovernmental Panel on Climate Change,* 2539–2654. Cambridge: Cambridge University Press, 2022.

IPCC. "Emissions Trends and Drivers." In *Climate Change 2022: Mitigation of Climate Change. Working Group III Contribution to the Sixth Assessment Report of the Intergovernmental Panel on Climate Change*, 215–294. Geneva: IPCC, 2022.

IPCC. "Point of Departure and Key Concepts." In *Climate Change 2022: Impacts, Adaptation and Vulnerability. Contribution of Working Group II to the Sixth Assessment Report of the Intergovernmental Panel on Climate Change,* 121–196. Cambridge: Cambridge University Press, 2022.

IPCC. "Summary for Policymakers." In *Climate Change 2022: Impacts, Adaptation and Vulnerability. Contribution of Working Group II to the Sixth Assessment Report of the Intergovernmental Panel on Climate Change,* 3–33. Cambridge: Cambridge University Press, 2022.

IPCC. "Summary for Policymakers." In *Climate Change 2023: Synthesis Report. A Report of the Intergovernmental Panel on Climate Change. Contribution of Working Groups I, II and III to the Sixth Assessment Report of the Intergovernmental Panel on Climate Change,* 1–34. Geneva: IPCC, 2023.

Isikara, Guney and Ozgur Narin. "Degrowth and Socialism: Notes on Some Critical Junctures." *Monthly Review* 75, no. 3 (2021): 30–43.

IUCN (International Union for Conservation of Nature). *Nature 2030: One Nature, One Future*. Marseilles: IUCN World Conservation Congress, 2021.

Izrael, Yu. A. "Climate Change Impact Studies: The IPCC Working Group II Report." In *Climate Change: Science, Impacts and Policy. Proceedings of the Second World Climate Conference,* edited by J. Jager and H. L. Ferguson. Cambridge: Cambridge University Press, 1991: 83–86.

Jackson, Peter. "From Stockholm to Kyoto: A Brief History of Climate Change." *UN Chronicle,* XLIV, no. 2, June 2007.

Jacobson, Chris, Stacy Crevello, Chanthan Chea, and Ben Jarihani. "When Is Migration a Maladaptive Response to Climate Change?" *Regional Environmental Change* 19, no. 10 (2019): 101–112.

Jamieson, Dale. *Reason in a Dark Time: Why the Struggle Against Climate Change Failed—and What It Means for Our Future*. New York: Oxford University Press, 2014.

Javeline, Debra. "The Most Important Topic Political Scientists Are Not Studying: Adapting to Climate Change." *Perspectives on Politics* 12, no. 2 (2014): 420–434.

Johnson, Alan R. "Avoiding Environmental Catastrophes: Varieties of Principled Precaution." *Ecology and Society* 17, no. 3 (2012), p. 9. Kahn, Matthew E. *Adapting to Climate Change: Markets and the Management of an Uncertain Future*. New Haven: Yale University Press, 2021.

Kanie, Norichika, Michele M. Betsill, Ruben Zondervan, Frank Biermann, and Oran R. Young. "A Charter Moment: Restructuring Governance for Sustainability." *Public Administration and Development* 32, no. 3 (2012): 292–304.

Kates, Robert W., William R. Travis, and Thomas J. Wilbanks. "Transformational Adaptation When Incremental Adaptations to Climate Change Are Insufficient." *Proceedings of the National Academy of Sciences of the United States of America* 109, no. 19 (2012): 7156–7161.

Kelly, Sean, and Joanna Macy. "The Great Turning: Reconnecting Through Collapse." In *Deep Adaptation: Navigating the Realities of Climate Chaos,* edited by Jem Bendell and Rupert Read, 227–237. Cambridge, England: Polity Press, 2021.

Keohane, Robert O. "The Global Politics of Climate Change: Challenge for Political Science." *PS: Political Science and Politics* 48, no. 1 (2015): 19–26.

Klein, Naomi. *This Changes Everything: Capitalism vs. the Climate*. New York: Simon and Schuster, 2014.

Kohona, Palitha. "Climate Change—Are We Really Confronting This Challenge?" *Environmental Policy and Law* 46, no. 2 (2016): 109–111.

Kolk, Ans, and Jonatan M. Pinske. "Multinationals' Political Activities on Climate Change." *Business and Society* 46, no. 2 (2007): 201–228.

Kovel, Joel. *The Enemy of Nature: The End of Capitalism or the End of the World?* London: Zed Books, 2007.

Krasner, Stephen, ed. *International Regimes*. Ithaca: Cornell University Press, 1983.

Kyle, Jessica. "Transformational Change Is Needed to Address Climate Change." ICF Climate Center, July 15, 2021.

Latouche, Serge. "Degrowth Economics." *Le Monde Diplomatique,* November 2004.

Latouche, Serge. *Essays on Frugal Abundance.* Westerville, Ohio: Simplicity Institute, 2014. https://simplicityinstitute.org.

Laybourn, Laurie, Henry Throw, and Suzannah Sherman. *1.5°C—Dead or Alive: The Risks to Transformational Change from Reaching and Breaching the Paris Agreement Goal.* London: Institute for Public Policy Research and Chatham House, 2023.

Lesnikowski, Alexandra C., James D. Ford, Lea Berrang-Ford, Magda Barrera, and Jody Heymann. "How Are We Adapting to Climate Change? A Global Assessment." *Mitigation and Adaptation Strategies for Global Change* 20, no. 2 (2015): 277–293.

Lesnikowski, Alexandra, James Ford, Robbert Biesbroek, Lea Berrang-Ford, Michelle Maillet, Malcolm Araos, and Stephanie E. Austin. "What Does the Paris Agreement Mean for Adaptation?" *Climate Policy* 17, no. 7 (2017): 825–831.

Lidskog, Rolf, and Goran Sundqvist. "Lost in Transformation: The Paris Agreement, the IPCC and the Quest for National Transformative Change." *Frontiers in Climate* 4, no. 906054 (2022). DOI 10.3389/fclim.2022.906054.

Lindsey, Rebecca, and Luann Dahlmann. "Climate Change: Ocean Heat Content." NOAA Climate.gov, August 17, 2020.

Liverman, Colin P., Summerhayes, Anthony D. Barnosky, Sarah E. Cornell, Michel Crucifix, Jonathan F. Donges, and Ingo Fetzer et al. "Trajectories of the Earth System in the Anthropocene." *Proceedings of the National Academy of Sciences* (2018):

Lomborg, Bjørn. *Global Warming—Are We Doing the Right Thing?* Aarhus: University of Aarhus, 2001.

Lomborg, Bjørn. "The Truth About the Environment." *The Economist,* August 2, 2001.

Lovelock, James. *The Revenge of Gaia*. New York: Basic Books, 2006.

Löwy, Michael. *Ecosocialism: A Radical Alternative to Capitalist Catastrophe*. Chicago: Haymarket Books, 2005.

Lowy, Michael. "Nine Theses on Ecosocialist Degrowth." *Monthly Review* 75, no. 3 (2023): 154–159.

Lütticken, Sven. "Unnatural History." *New Left Review* 45 (2007): 115–131.

Lynas, Mark. *Our Final Warning: Six Degrees of Climate Emergency*. London: 4th Estate, 2020.

Macron, Emmanuel, Mia Amor Mottley, Luiz Inacio Lula da Silva, Ursula von der Leyen, Charles Michel, Olaf Scholz, and Fumio Kishida et al. "A Green Transition that Leaves No One Behind." *Project Syndicate*, June 20, 2023.

Magdoff, Fred. "Ecological Civilization." *Monthly Review* 62, no. 8 (2011): 1–25.

Magdoff, Fred. "Harmony and Ecological Civilization: Beyond the Capitalist Alienation of Nature." *Monthly Review* 64, no. 2 (2012): 1–9.

Magdoff, Fred, and John Bellamy Foster. "What Every Environmentalist Needs to Know About Capitalism." *Monthly Review* 61, no. 10 (2010): 1–30.

Magdoff, Fred, and Chris Williams. *Creating an Ecological Society: Toward a Revolutionary Transformation*. New York: Monthly Review Press, 2017.

Magnan, Alexandre. "Avoiding Maladaptation to Climate Change: Towards Guiding Principles." *S.A.P.I.EN.S.* 7, no. 1 (2014):

Magnus, Benzie. "Social Justice and Adaptation in the UK." *Ecology and Society* 19, no. 1 (2014), p. 39.

Marcott, Shaun A., Jeremy D. Shakun, Peter U. Clark, and Alan C. Mix. "A Reconstruction of Regional and Global Temperature for the Past 11,300 Years." *Science* 339, no. 6124 (2013): 1198–1201.

Martin, Susan. "Climate Change, Migration, and Governance." *Global Governance* 16, no. 3 (2010): 397–414.

Marx, Karl, and Friedrich Engels. "The Communist Manifesto (1848)." In *The Portable Marx*, edited by Eugene Kamenka, 203–241. New York: Penguin Books, 1983.

Matsuoka, Shunji. "The Genesis and End of Institutional Fragmentation in Global Governance on Climate Change from a Constructivist Perspective." *International Environmental Agreements: Politics, Law and Economics* 17, no. 2 (2017): 143–159.

McAlpine, Clive A., Leonie M. Seabrook, Justin G. Ryan, Brian J. Feeney, William J. Ripple, Anne H. Ehrlich, and Paul R. Ehrlich. "Transformational Change: Creating a Safe Operating Space for Humanity." *Ecology and Society* 20, no. 1 (2015), p. 56.

McKibben, Bill. *Eaarth: Making Life on a Tough New Planet*. New York: Henry Holt, 2010.

McKibben, Bill. *Falter: Has the Human Game Begun to Play Itself Out?* New York: Henry Holt, 2019.

McLaughlin, Paul. "Climate Change, Adaptation, and Vulnerability: Reconceptualizing Societal-Environment Interaction Within a Socially Constructed Adaptive Landscape." *Organization and Environment* 24, no. 3 (2011): 269–291.

McMichael, Anthony J., and Keith B. Dear. "Climate Change: Heat, Health, and Longer Horizons." *Proceedings of the National Academy of Sciences* 107, no. 21 (2010): 9483–9484.

Meadows, Donella, Dennis Meadows, Jørgen Randers, and William Behrens III. *The Limits to Growth: A Report for the Club of Rome's Project on the Predicament of Mankind*. New York: Signet, 1972.

Meszaros, Istvan. "The Only Viable Economy." *Monthly Review* 58, no. 11 (2007): 13–33.

Methmann, Chris, and Delf Rothe. "Politics for the Day After Tomorrow: The Logic of Apocalypse in Global Climate Politics." *Security Dialogue* 43, no. 4 (2012): 323–344.

Mintzer, Irving. "Communication on Energy: Is the Heat On?" *Energy Policy*16, no. 1 (1988): 69–72.

Mintzer, Irving. "Living in a Warmer World: Challenges for Policy Analysis." *Journal of Policy Analysis and Management* 7, no. 3 (1988): 445–459.

Möhner, Annett, and Richard J. T. Klein. *The Global Environment Facility: Funding for Adaptation or Adapting to Funds?* Stockholm: Stockholm Environment Institute, 2007.

Moore, Frances C. "Costing Adaptation: Revealing Tensions in the Normative Basis of Adaptation Policy in Adaptation Cost Estimates." *Science, Technology, and Human Values* 37, no. 2 (2012): 171–198.

Moore, Jason W. "Ecology, Capital, and the Nature of Our Times: Accumulation & Crisis in the Capitalist World-Ecology." *Journal of World-Systems Research* 17, no. 1 (2011): 107–146.

Moore, Jason W. *Capitalism in the Web of Life: Ecology and the Accumulation of Capital.* London: Verso, 2015.

Moore, Michele-Lee, Ola Tjornbo, Elin Enfors, Corrie Knapp, Jennifer Hodbad, Jacopo A. Baggio, Albert Norstrom, Per Olsson, and Duan Biggs. "Studying the Complexity of Change: Toward an Analytical Framework for Understanding Deliberate Social-Ecological Transformations." *Ecology and Society* 19, no. 4 (2014), p. 54.

Morton, Oliver. "The Geopolitical Challenges of Geoengineering—and Geoengineering's Challenge to Geopolitics." Washington, DC: Wilson Center, September 30, 2020.

Moser, Susanne C., and Julia A. Ekstrom. "A Framework to Diagnose Barriers to Climate Change Adaptation." *Proceedings of the National Academy of Sciences of the United States of America* 107, no. 51 (2010): 22026–22031.

Mustelin, Johanna, and John Handmer. "Triggering Transformation: Managing Resilience or Invoking Real Change?" In *University of Oslo, Proceedings: Transformation in a Changing Climate, International Conference in Oslo, Norway (June 19–21),* edited by Karen O'Brien, Linda Sygna, Asuncion Lera St. Clair, Per Olsson, Heide Hackmann, and Eva Bakkeslett, 24–32. Oslo: University of Oslo, 2013.

National Research Council. *Adapting to the Impacts of Climate Change.* Washington, DC: National Academies Press, 2010.

New, Mark, Diana Liverman, Heike Schroder, and Kevin Anderson. "Introduction: Four Degrees and Beyond: The Potential for a Global Temperature Increase of Four Degrees and Its Implications." *Philosophical Transactions: Mathematical, Physical and Engineering Sciences* 369, no. 1934 (2011): 6–19.

Newell, Peter. "The Elephant in the Room: Capitalism and Global Environmental Change." *Global Environmental Change* 21, no. 1 (2011): 4–6.

Newell, Peter, Shilpi Srivastava, Lars Otto Naess, Gerardo A. Torrest, and Roz Price. "Toward Transformative Climate Justice: Key Challenges and Future Directions for Research." *Institute of Development Studies Working Papers 2020* 540 (2020).

Newitz, Annalee. *Scatter, Adapt, and Remember: How Humans Will Survive a Mass Extinction.* New York: Doubleday, 2013.

Nitzan, Jonathan, and Shimshon Bichler. *Capital as Power.* RIPE Series in Global Political Economy. New York: Routledge, 2009.

Nordhaus, Ted, Vijaya Ramachandran, and Patrick Brown. "The Obvious Climate Strategy Nobody Will Talk About: Economic Development Is the Only Proven Path to Climate Resilience," *Foreign Policy,* November 6, 2022.

Nordhaus, William D. "A Perspective on Costs and Benefits." *EPA Journal* 16, no. 2 (1990): 44–45.

Nordhaus, William D. "To Slow or Not to Slow: The Economics of the Greenhouse Effect." *The Economic Journal* 101, no. 407 (1991): 920–937.

Norstrom, Albert V., Astrid Dannenberg, Geoff McCarney, Manjana Milkoreit, Florian Diekert, Gustav Engstrom, and Ram Fishman et al. "Three Necessary Conditions for Establishing Effective Sustainable Development Goals in the Anthropocene." *Ecology and Society* 19, no. 3 (2014), p. 8.

O'Brien, Karen. "Global Environmental Change II: From Adaptation to Deliberate Transformation." *Progress in Human Geography* 36, no. 5 (2012): 667–676.

O'Brien, Karen L., and Elin Selboe, eds. *The Adaptive Challenge of Climate Change*. New York: Cambridge University Press, 2015.

O'Brien, Karen, and Linda Sygna. "Responding to Climate Change: The Three Spheres of Transformation." in *University of Oslo, Proceedings: Transformation in a Changing Climate, International Conference in Oslo, Norway (June 19–21),* edited by Karen O'Brien, Linda Sygna, Asuncion Lera St. Clair, Per Olsson, Heide Hackmann, and Eva Bakkeslett, 16–23. Oslo: University of Oslo, 2013.

O'Brien, Karen, Linda Sygna, Asuncion Lera St. Clair, Per Olsson, Heide Hackmann and Eva Bakkeslett. "The First Transformation in a Changing Climate Conference: Introduction and Reflections." In *University of Oslo, Proceedings: Transformation in a Changing Climate, International Conference (June 19–21),* edited by Karen O'Brien, Linda Sygna, Asuncion Lera St. Clair, Per Olsson, Heide Hackmann, and Eva Bakkeslett, 4–7. Oslo: University of Oslo, 2013.

O'Connor, James. "A Prolegomenon to an Ecological Marxism: Thoughts on the Materialist Conception of History." *Capitalism Nature Socialism* 10, no. 2 (1999): 77–106.

Olsen, Karen H., and Neelam Singh, eds. *Transformational Change Methodology: Assessing the Transformational Impacts of Policies and Actions*. Washington, DC: World Resources Institute, 2020.

Ord, Toby. *The Precipice: Existential Risk and the Future of Humanity*. New York: Hachette Books, 2020.

Orton, David. "Reflections on Rio." *Canadian Dimension* 26, no. 7 (1992): 17.

Osborn, Jen Flatt. "Forests Under Threat: A Comprehensive Look at the Latest Deforestation Statistics." World Animal Foundation, April 7, 2023. https://worldanimalfoundation.org

Ostrom, Elinor. *Governing the Commons: The Evolution of Institutions for Collective Action.* Cambridge: Cambridge University Press, 1990.

Park, S. E., N. Marshall, E. Jakku, A. Dowd, S. Howden, E. Mendham, and A. Fleming. "Informing Adaptation Responses to Climate Change Through Theories of Transformation." *Global Environmental Change* 22 (2012): 115–126.

Parry, Jo-Ellen. *Building a Climate-Resilient City: Transformational Adaptation.* Winnipeg: IISD and the University of Winnipeg, 2017.

Parsons, Laurie. *Carbon Colonialism: How Rich Countries Export Climate Breakdown.* Manchester: Manchester University Press, 2023.

Parsons, Meg, and Johanna Nalau. "Historical Analogies as Tools in Understanding Transformation." *Global Environmental Change* 38 (2016): 82–96.

Paterson, Matthew. *Global Warming and Global Politics*. London: Routledge, 1996.

Paterson, Matthew. "Climate Change and International Political Economy: Between Collapse and Transformation." *Review of International Political Economy* 28, no. 2 (2021): 394–405.

Paterson, Matthew, and Michael Grubb. "The International Politics of Climate Change." *International Affairs* 68, no. 2 (1992): 293–310.

Pelling, Mark. *Adaptation to Climate Change: From Resilience to Transformation.* London: Routledge, 2011.

Pelling, Mark. "Resilience and Transformation." In *Climate Change and the Crisis of Capitalism: A Chance to Reclaim, Self, Society and Nature,* edited by Mark Pelling, Daniel Manuel-Navarette, and Michael Redclift, 51–65. London: Taylor and Francis, 2011.

Pelling, Mark, Karen O'Brien, and David Matyas. "Adaptation and Transformation." *Climatic Change* 133, no. 1 (2015): 113–127.

Pepper, David. "Sustainable Development and Ecological Modernization: A Radical Homocentric Perspective." *Sustainable Development* 6, no. 1 (1998): 1–7.

Pickering, Jonathan, Thomas Hickmann, Karin Backstrand, Agni Kalfagianni, Michael Bloomfield, Aysem Mert, Hedda Ransan-Cooper, and Alex Y. Lo. "Democratizing Sustainability Transformations: Assessing the Transformative Potential of Democratic Practices in Environmental Governance." *Earth System Governance* 11 (2022): 100131.

Pielke, Roger A., Jr. "The Case for a Sustainable Climate Policy: Why Costs and Benefits Must Be Temporally Balanced." *University of Pennsylvania Law Review* 155, no. 6 (2007): 1843–1857.

Pielke, Roger A., Jr., Gwyn Prins, Steve Rayner, and Daniel Sarewitz. "Lifting the Taboo on Adaptation." *Nature* 445, no. 7128 (2007): 597–598.

Rasmus, Jack. *The Scourge of Neoliberalism: US Economic Policy from Reagan to Trump.* Atlanta: Clarity Press, 2020.

Raworth, Kate. *A Safe and Just Space for Humanity: Can We Live Within the Doughnut?* Oxford: Oxfam, 2012.

Read, Rupert. "This Civilization Is Finished." *The Ecological Citizen* 3, no. 2 (2020): 157–162.

Read, Rupert. "Riding Two Horses: The Future of Politics and Activism, as We Face Potential Eco-Driven Societal Collapse." In *Deep Adaptation: Navigating the Realities of Climate Chaos,* edited by Jem Bendell and Rupert Read, 274–298. Cambridge, England: Polity Press, 2021.

Ribot, Jesse. "Vulnerability Before Adaptation: Toward Transformative Climate Action." *Global Environmental Change* 21, no. 4 (2011): 1160–1162.

Rickards, Lauren, and Stuart Howden. "Transformational Adaptation: Agriculture and Climate Change." *Crop and Pasture Science* 63, no. 3 (2012): 240–250.

Ripple, William J., Christopher Wolf, Thomas M. Newsome, Phoebe Barnard, and William R. Moomaw. "World Scientists' Warning of a Climate Emergency." *BioScience* 70, no. 1 (2020): 8–12.

Roberts, Leslie. "Academy Panel Split on Greenhouse Adaptation: Its Conclusion that the United States Can Adapt Relatively Painlessly to Global Warming Draws Two Vigorous Dissents." *Science* 253, no. 5025 (1991), p. 1206.

Roche, David. *The United Nations in the 21st Century.* Toronto, Ontario, Canada: James Lorimer, 2015.

Rockström, Johan. "A Resilient Earth for Future Generations." In *Living Planet Report 2016: Risk and Resilience in a New Era,* edited by World Wide Fund for Nature, 4–5. Gland: WWF International, 2016.

Rockstrom, Johan, Will Steffen, Kevin Noone, Asa Persson, F. StuartChapin III, Eric Lambin, and Timothy M. Lenton et al. "Planetary Boundaries: Exploring the Safe Operating Space for Humanity." *Ecology and Society* 14, no. 2 (2009), p. 32.

Roggema, Rob, Tim Vermeend, and Andy Van den Dobbelsteen. "Incremental Change, Transition or Transformation? Optimising Change Pathways for Climate Adaptation in Spatial Planning." *Sustainability* 4, no. 10 (2012): 2525–2549.

Rothe, Delf. "Managing Climate Risks or Risking a Managerial Climate: State, Security and Governance in the International Climate Regime." *International Relations* 25, no. 3 (2011): 330–345.

Saran, Shyam. "Global Governance and Climate Change." *Global Governance* 15, no. 4 (2009): 457–460.

Sarma, K. Madhava. "Adaptation Measures." In *Climate Change: Science, Impacts and Policy. Proceedings of the Second World Climate Conference,* 135–138. Edited by Jill Jager and Howard L. Ferguson. Cambridge: Cambridge University Press, 1991.

Schlosberg, David. "Political Challenges of the Climate-Changed Society." *PS: Political Science and Politics* 46, no. 1 (2013): 13–17.

Schneider, Francois, Joan Martinez-Alier, and Giorgio Kallis. "Sustainable Degrowth." *Journal of Industrial Ecology* 15, no. 5 (2011): 654–656.

Schultheiß, Lisa, Zoha Shawoo, Inès Bakhtaoui, Lina Ahmed, Courtney Lindsay, and Arunima Sircar. *Operationalising the Loss and Damage Fund: Learning from the Funding Mosaic.* Bonn: Germanwatch e.V., 2023.

Schweickart, David. *After Capitalism*. Lanham, MD: Rowman and Littlefield, 2002.

Schweickart, David. "Is Sustainable Capitalism Possible?" *Procedia Social and Behavioral Sciences* 2, no. 5 (2010): 6739–6752.

Scoones, Ian, Andrew Stirling, Dinesh Abrol, Joanes Atela, Lakshmi Charli-Joseph, Hallie Eakin, and Adrian Ely et al. "Transformations to Sustainability: Combining Structural, Systemic and Enabling Approaches." *Current Opinion in Environmental Sustainability* 42 (2020): 65–75.

Servigne, Pablo, Raphael Stevens, Gauthier Chapelle, and Daniel Rodary. "Reasons for Anticipating Societal Collapse." 108–125. In *Deep Adaptation: Navigating the Realities of Climate Chaos,* edited by Jem Bendell and Rupert Read. Cambridge: Polity Press, 2021.

Sewell, William H., Jr. *Logics of History: Social Theory and Social Transformation.* Chicago; University of Chicago Press, 2005.

Sherwood, Steven C., and Matthew Huber. "An Adaptability Limit to Climate Change Due to Heat Stress." *Proceedings of the National Academy of Sciences* 107, no. 21 (2010): 9552–9555.

Sinha, Suresh K. "Impact of Climate Change on Agriculture: A Critical Assessment." In *Climate Change: Science, Impacts and Policy. Proceedings of the Seecond World Climate Conference*, edited by Jill Jager and Howard L. Ferguson, 99–107. Geneva: Secretariat of the World Meteorological Organization, 1979.

Sklair, Leslie. "Social Movements for Global Capitalism: The Transnational Capitalist Class in Action." *Review of International Political Economy* 4, no. 3 (1997): 514–538.

Sklair, Leslie. "World Revolution or Socialism, Community by Community, in the Anthropocene?" *Journal of World-Systems Research* 25, no. 2 (2019): 297–305.

Sklair, Leslie. *The Anthropocene in Global Media: Neutralizing the Risk.* Routledge Studies in Environmental Communication and Media. Abingdon: Routledge, 2021.

Slater, Matthew, and Skeena Rathor. "Relocalization as Deep Adaptation." In *Deep Adaptation: Navigating the Realities of Climate Chaos,* edited by Jem Bendell and Rupert Read, 299–324. Cambridge, England: Polity Press, 2021.

Slaughter, Richard A. "Evaluating 'Overshoot and Collapse' Futures." *World Future Review* 2, no. 4 (2010): 5–18.

Smith, Mark Stafford. "Responding to Global Environmental Change." In *Change! Combining Analytic Approaches With Street Wisdom,* edited by Gabriele Bammer, 29–42. Canberra: ANU Press, 2015.

Sovacool, Benjamin K., and Björn-Ola Linnér. *The Political Economy of Climate Change Adaptation.* New York: Palgrave Macmillan, 2016.

Spratt, David. *Dangerous Climate Change: Myths and Reality*. Melbourne: Climate Code Red, 2014.

Spratt, David, and Ian Dunlop. *What Lies Beneath: The Underestimation of Existential Climate Risk.* Melbourne, Australia: National Centre for Climate Restoration, 2018.

Sprinz, Detlef, and Tapani Vaahtoranta. "The Interest-Based Explanation of International Environmental Policy." *International Organization* 48, no. 1 (1994): 77–105.

Steffen, Will, Johan Rockstrom, Katherine Richardson, Timothy M. Lenton, Carl Folke, Diana Liverman, and Colin P. Summerhayes et al. "Trajectories of the Earth System in the Anthropocene." *Proceedings of the National Academy of Sciences* 115 no. 33 (2018): 8252–8259. DOI 10.1073/pnas.1810141115.

Sygna, Linda, Karen O'Brien, and Johanna Wolf, eds. *A Changing Environment for Human Security: Transformative Approaches to Research, Policy and Action.* London: Routledge, 2013.

Tanner, Thomas M., and Aditya V. Bahadur. "Distilling the Characteristics of Transformational Change in a Changing Climate." In *University of Oslo, Proceedings: Transformation in a Changing Climate, International Conference in Oslo, Norway (June 19–21),* edited by Karen O'Brien, Linda Sygna, Asuncion Lera St. Clair, Per Olsson, Heide Hackmann, and Eva Bakkeslett, 33–42. Oslo: University of Oslo, 2013.

Termeer, Catrien J. A. M., Art Dewulf, and G. Robbert Biesbroek. "Transformational Change: Governance Interventions for Climate Change Adaptation from a Continuous Change Perspective." *Journal of Environmental Planning and Management* 60, no. 4 (2017): 558–576.

Thaler, Mathias. "Eco-Miserablism and Radical Hope: On the Utopian Vision of Post-Apocalyptic Environmentalism." *American Political Science Review*188, no. 1 (2023): 318–331.

Tokar, Brian. "The Paris Climate Agreement: Hope or Hype?" ZNET, December 30, 2015. https://zcomm.org

Tschakert, Petra, and Asuncion Lera St. Clair. "Conditions for Transformative Change: The Role of Responsibility, Care, and Place-Making in Climate Change Research." In *University of Oslo, Proceedings of Transformation in a Changing Climate 19–21 June 2013.* edited by Karen O'Brien, Linda Sygna, Asuncion Lera St. Clair, Per Olsson, Heide Hackmann, and Eva Bakkeslett, 276–275. Oslo: University of Oslo, 2013.

Turner, Billie L., II "Vulnerability and Resilience: Coalescing or Paralleling Approaches for Sustainability Science." *Global Environmental Change* 20, no. 4 (2010): 570–576.

UN Adaptation Committee. *25 Years of Adaptation Under the UNFCCC*. Bonn: UN Climate Change Secretariat, 2019.

UN Adaptation Fund Board. *Draft Provisional Operational Policies and Guidelines for Parties to Access Resources from the Adaptation Fund.* AFB/B/3/8, 2008.

UNCED (UN Conference on Environment and Development). *Agenda 21: Programme of Action for Sustainable Development, Rio Declaration on Environment and Development, Statement of Forest Principles: The Final Text of Agreements Negotiated by Governments at the United Nations Conference on Environment and Development*. June 3–14, 1992, Rio de Janeiro, Brazil.

UNEP (UN Environment Programme). *Adaptation Gap Report 2020*. Nairobi: UNEP, 2021.

UNEP (UN Environment Programme). *Adaptation Gap Report 2021: The Gathering Storm—Adapting to Climate Change in a Post-Pandemic World—Executive Summary*. Nairobi: UNEP, 2021.

UNEP. *Adaptation Gap Report 2022: Too Little, Too Slow—Climate Adaptation Failure Puts World at Risk*. Nairobi: UNEP, 2022.

UNEP. *Adaptation Gap Report 2023: Underfinanced and Underprepared. Inadequate Investment and Planning on Climate Adaptation Leaves World Exposed*. Nairobi: UNEP, 2023.

UNFCCC (UN Framework Convention on Climate Change). *Report of the Conference of the Parties on Its Seventh Session, Held at Marrakesh from 29 October to 10 November 2001, Addendum, Part Two, Actions Taken by the Conference of the Parties*. FCCC/CP/2001/13/Add.1, 2002.

UNFCCC. *Report of the Conference of the Parties on Its Thirteenth Session, Held in Bali from 3 to 15 December 2007, Addendum Part Two: Action Taken by the Conference of the Parties at Its Thirteenth Session*, FCCC/CP/2007/6/Add.1, Decision 1/CP.13, 2008.

UNFCCC. *Report of the Conference of the Parties on Its Sixteenth Session, Held in Cancun from 29 November to 10 December 2010, Addendum Part Two: Action Taken by the Conference of the Parties at Its Sixteenth Session*. FCCC/CP/2010/7/Add.1, 2010.

UNFCCC. *Workshops Under the Glasgow-Sharm el-Sheikh Work Programme on the Global Goal on Adaptation*. Report by the Secretariat. FCCC/SB/2022/INF.2, 2022.

UNFCCC. *Summary of the Fifth Workshop Under the Glasgow–Sharm El-Sheikh Work Programme on the Global Goal on Adaptation: Changing Mindsets and Worldviews Towards Transformation in Adaptation, with the Inclusion of Indigenous Peoples' Wisdom, Values and Knowledge and Consideration of Cross-Cutting Issues Included in Paragraph 20 (G) (E) of Decision 3/cma.4*. 2023. https://unfccc.int.

UNFCCC. *Technical Dialogue of the First Global Stocktake*. FCCC/SB/2023/9, Advance Version, September 8, 2023.

UNFCCC. *Glasgow–Sharm el-Sheikh Work Programme on the Global Goal on Adaptation*. n.d. https://unfccc.int.

USGCRP (US Global Change Research Program). "Reducing Risks Through Adaptation Action." In *Impacts, Risks, and Adaptation in the United States: Fourth National Climate Assessment*, vol. 2, 1309–1345. Washington, DC: USGCRP, 2018. DOI 10.7930/NCA4.2018.

van de Sand, Isabel. "Payments for Ecosystem Services in the Context of Adaptation to Climate Change." *Ecology and Society* 17, no. 1 (2012), p. 11.

Vanderheiden, Steve. "Coaxing Climate Policy Leadership." *Ethics and International Affairs* 26, no. 4 (2012): 463–479.

Vanderheiden, Steve. "Introduction." In *Political Theory and Global Climate Change*, edited by Steve Vanderheiden, xi–xxiv. Cambridge: MIT Press, 2008.

Veldman, Robin Globus. "Narrating the Environmental Apocalypse: How Imagining the End Facilitates Moral Reasoning Among Environmental Activists." *Ethics and the Environment* 17, no. 1 (2012): 1–23.

Vinke, Kira, Jonas Bergmann, Julia Blocher, Himani Upadhyay, and Roman Hoffmann. "Migration as Adaptation?" *Migration Studies* 8, no. 4 (2020): 626–634.

Waltz, Kenneth N. *Theory of International Politics*. Reading, MA: Addison-Wesley, 1979.

Weihe, Wolf H., and Raf Mertins. "Human Well-Being, Diseases and Climate." In *Climate Change: Science, Impacts and Policy. Proceedings of the Second World Climate Conference,* edited by Jill Jager and Howard. L. Ferguson, 345–359. Cambridge: Cambridge University Press, 1991.

Wendt, Alexander. *Social Theory of International Politics*. Cambridge: Cambridge University Press, 1999.

White, Damian F., Alan P. Rudy, and Brian J. Gareau. *Environments, Natures and Social Theory: Towards a Critical Hybridity*. London: Palgrave Macmillan, 2016.

White, Robert M. "Climate at the Millennium: Keynote Address." In *Proceedings of the World Climate Conference: A Conference of Experts on Climate and Mankind,* World Meteorological Association. Geneva: Secretariat of the World Meteorological Organization, 1979.

Williams, Chris. *Ecology and Socialism*. Chicago: Haymarket Books, 2010.

Williams, Chris. "How Will We Get to an Ecological Civilization?" *Climate and Capitalism,* November 2, 2015.

Williston, Byron. "Climate Change and Radical Hope." *Ethics and the Environment* 17, no. 2 (2012): 166–186.

Wilson, Samuel, Leonie J. Pearson, Yoshihisa Kashima, Dean Lusher and Craig Pearson. "Separating Adaptive Maintenance (Resilience) and Transformative Capacity of Social-Ecological System." *Ecology and Society* 18, no. 1 (2013), p. 22.

Wise, Russ M., Ioan Fazey, Mark Stafford Smith, Sarah E. Park, Hallie C. Eakin, Emma R. M. Archer Van Garderen, and Bruce M. Campbell. "Reconceptualising Adaptation to Climate Change as Part of Pathways of Change and Response." *Global Environmental Change* 28 (2014): 325–336.

WMO (World Meteorological Organization). *Proceedings of the World Climate Conference: A Conference of Experts on Climate and Mankind*. Geneva: Secretariat of the World Meteorological Organization, 1979.

WMO. *World Climate Conference: Extended Summaries of Papers Presented at the Conference*. Geneva: Secretariat of the World Meteorological Organization, 1979.

WMO. *Report of the International Conference on the Assessment of the Role of Carbon Dioxide and of Other Greenhouse Gases in Climate Variations and Associated Impacts*. Geneva: World Climate Programme, 1985.

Wolf, Johanna, Karen O'Brien, and Linda Sygna. "The Future Is Now." In *A Changing Environment for Human Security: Transformative Approaches to Research, Policy and Action,* edited by Linda Sygna, Karen O'Brien, and Johanna Wolf., p. 38. London: Routledge, 2013.

Wolff, Richard. *Democracy at Work: A Cure for Capitalism*. Chicago: Haymarket Books, 2012.

WRI (World Resources Institute), in collaboration with UNDP (UN Development Programme), UNEP (UN Environment Programme), and the World Bank. *World Resources 2010–2011: Decision Making in a Changing Climate—Adaptation Challenges and Choices.* Washington, DC: WRI, 2011.

York, Richard, Eugene A. Rosa, and Thomas Dietz. "Ecological Modernization Theory: Theoretical and Empirical Challenges." In *The International Handbook of*

Environmental Sociology, 2nd ed., edited by Michael R. Redclift and Graham Woodgate, 77–90. Cheltenham: Edward Elgar, 2010.

Yuen, Eddie. "The Politics of Failure Have Failed: The Environmental Movement and Catastrophism." In *Catastrophism, the Apocalyptic Politics of Collapse and Rebirth,* edited by David McNally Sasha Lilley, Eddie Yuen, and James Davis, 15–43. Toronto, Ontario, Canada: Between the Lines Books, 2012.

Zacher, Mark W., and Richard A. Matthew. "Liberal International Theory: Common Threads, Divergent Strands." In *Controversies in International Relations Theory: Realism and the Neoliberal Challenge,* edited by Charles W. Kegley Jr., 107–150. New York: St. Martin's Press, 1995.

Zehr, Stephen. "The Sociology of Global Climate Change." *WIREs Climate Change* 6, no. 2 (2015): 129–150.

Zillman, John W. "A History of Climate Activities." *Bulletin* 58, no. 3 (2009): 141–150.

Index

About the Book

HOW TO COPE WITH CLIMATE CHANGE? OBSERVING THAT efforts to mitigate rising temperatures are falling disastrously short, John Barkdull argues that policy must shift toward adaptation and considers what this entails. As he assesses climate policies and politics since the landmark 1992 Rio conference, Barkdull explores approaches to transformational adaptation that will allow us to survive in a much warmer world.

John Barkdull is associate professor emeritus of political science and director of global studies at Texas Tech University.